A NATURALIST'S

BUTTE

OF

MALAYSIA

Peninsular Malaysia, Singapore and Southern Thailand

Laurence Kirton

JOHN BEAUFOY PUBLISHING

Reprinted in 2018
First published in the United Kingdom in 2014 by John Beaufoy Publishing Limited
11 Blenheim Court, 316 Woodstock Road, Oxford OX2 7NS, England
www.johnbeaufoy.com

In association with the Forest Research Institute Malaysia
52109 Kepong, Selangor, Malaysia

10 9 8 7 6 5 4 3 2 1

Cover and Prelim Photos
Front cover: *main image* Painted Jezebel (Khew Sin Khoon); *bottom left* Great Egg-fly (Tan Chung Pheng); *bottom centre* Ashy-white Tree-Nymph (Sunny Chir); *bottom right* Great Mormon (Anthony Wong). **Back cover:** Rajah Brooke's Birdwing (Sunny Chir). **Title page:** Glorious Begum (Goh Lai Chong). **Contents page:** Red Harlequin (Sunny Chir).

Photographic Contributors
Anthony Wong, Antonio Giudici, Benedict Tay, Benjamin Yam, Bobby Mun, Chng Chuen Kiong, Dave Sargeant, Ellen Tan, Federick Ho, Games Punjapa Phetsri, Gan Cheong Weei, Glenn Q. Bagnas, Goh Lai Chong, Guek Hock Ping, Hiroyuki Hashimoto, Horace Tan, James Chia, John Moore, Jonathan Soong, Khew Sin Khoon, Koh Cher Hern, Laurence G. Kirton, Leslie Day, Liew Nyok Lin, Loke Peng Fai, Mark Wong, Mohd Walad Jamaludin, Nawang Bhutia, Nelson Ong, Oleg Kosterin, Peter Eeles, Richard Ong, Simon Sng, Steeve Collard, Sum Chee Meng, Sunny Chir, Tan Ben Jin, Tan Chung Pheng, Tarun Karmakar, Tea Yi Kai, Terry Ong and Thawat Tanhai.
Full photo credits for species descriptions can be found on p.172.

ISBN 978-1-912081-62-2

Edited and designed by D & N Publishing, Baydon, Wiltshire, UK
Printed and bound in Malaysia by Times Offset (M) Sdn. Bhd.

Acknowledgements
Many butterfly enthusiasts generously contributed photographs and are listed above. Khew Sin Khoon and Leslie Day, in addition, helped provide useful contacts and source some of the photographs used in this book. Seow Tek Lin provided advice and concurrence on the identities of butterflies in some of the photographs. Chooi-Khim Phon helped in various ways, including in preparing the classification of butterfly genera in the region. Adam Cotton provided updated information on the taxonomy of some swallowtail butterflies, and together with Dave Sargeant and Leslie Day, provided useful information on the behaviour and habitats of some butterfly species found only in Thailand. Henry Barlow and Rienk de Jong reviewed the first draft and provided useful advice. Thanks go to each of these persons who contributed in their own way to the production of this book … all for the love of butterflies. In addition, thanks go to James Maxwell for supplying useful articles on which the account of Thailand's vegetation is based.

·Contents·

▪ Introduction, Geography & Climate ▪

Introduction

About 1,400 species of butterflies are known to occur in the region comprising Thailand, Peninsular Malaysia and Singapore. This field guide covers 470 butterfly species, of which 280 are illustrated. They have been selected to include the most commonly encountered species and to have as good a representation as possible of different butterfly groups. Where butterflies belong to a diverse group, this is mentioned in the text in order to give readers a feeling for the breadth of species that may resemble the species covered.

Wherever possible, characters that are most easily seen in living butterflies have been used, and their description has been simplified to make them easier to use in the field. Forewing length has been used as an indication of size rather than wingspan because it is less ambiguous and easier to conceptualise in the field.

Geography

Thailand extends from the continent of Asia southwards into the sea through a narrow land bridge known as the Isthmus of Kra. Further south is West or Peninsular Malaysia, with the island nation of Singapore at its southern tip. Peninsular Malaysia is divided into states and Thailand is divided into many official provinces. Thailand also has six unofficial regions: North, Northeast, Central, East, West and South (or Peninsular) Thailand. Peninsular Malaysia, Singapore and Thailand are in the heart of the Oriental Realm, a biogeographical area stretching from the east of Pakistan through India, to South China and Southeast Asia as far east as the Philippines, Sulawesi and Timor. As a result, they share many species in common. Singapore and most of Peninsular Malaysia, together with Sumatra, Java, Bali, Borneo and Palawan, lie in a biogeographical region called the Sundanian Region or Sundaland, which was once a large connected landmass during the ice ages when sea levels were lower. Thailand is in the centre of the biogeographical region known as Indo-Burma (sometimes called the Indo-Chinese Region), which comprises the north-eastern extreme of India east of the Himalayan range, Burma, Thailand, Laos, Cambodia, Vietnam and South China.

Climate

Thailand, Peninsular Malaysia and Singapore lie north of the equator and have a climate that can be broadly described as tropical or equatorial. In Thailand, the average annual temperature ranges from 19 to 38°C in the lowlands. Peninsular Malaysia has more uniform temperatures ranging between 23 and 32°C. At higher altitudes, average temperatures drop. Singapore has about the same average temperatures as the Peninsula but lacks the low temperatures of the highlands.

The average annual rainfall in Thailand ranges from 1,020mm in the northeast to over 4,000mm in parts of the Thai peninsula. In Peninsular Malaysia average annual rainfall ranges from about 1,750mm in the northwest to about 6,500mm in the northeast, with a national average of about 2,800mm. Singapore receives 2,340mm of rainfall on average per year. Humidity in the Peninsula and Singapore tends to be high, between 70 to 90%.

Two monsoon winds blow in the region and have a strong influence on rainy seasons. The southwest monsoon winds blow from the Indian Ocean from about May to October. Between August and September the winds reverse and the northeast monsoon blows in from the South China Sea till about March. The duration and onset of rainy seasons and the amount of rain vary with latitude, altitude and the presence of mountain ranges that create rain shadows.

Continental Thailand has, more specifically, a tropical wet-and-dry climate (or savannah climate) with three distinct seasons. The rainy season occurs from June to October, often with short, heavy rain in the afternoon. It is followed by a cool but relatively dry season from November to February. The hot, dry season occurs from March to May. The northeast experiences a longer dry season.

Peninsular Thailand and the southeastern tip of Continental Thailand have a tropical monsoon climate with two seasons: a rainy season from June to December, and a hot, dry season from January to May. Peninsular Malaysia has a tropical monsoon climate only at its northwestern end. Here the southwest monsoon brings a rainy season from August to October, while the northeast monsoon brings a dry season from December to February because its moisture is shed over the central mountain ranges.

In the rest of Peninsular Malaysia as well as in Singapore, the climate is equatorial with abundant rainfall throughout the year and less distinct seasons. The wettest periods west of the peninsula's central mountain ranges occur from October to November and April to May. The wettest periods on the east are from November to January. The driest periods in most of the Peninsula and Singapore are from June to July, and the month of February.

Unusually dry or wet years may occur as a result of the El Niño/La Niña–Southern Oscillation (ENSO) effect. Other extreme weather events are uncommon in Thailand, Peninsular Malaysia and Singapore. Thailand is occasionally affected by typhoons; Malaysia and Singapore are not but may have unusually wet weather when typhoons and tropical storms affect neighbouring regions.

VEGETATION

The main inland forest formations of the region can be broadly classified as tropical rainforests and tropical seasonal forests. In addition, there are various other specialised coastal and inland vegetation types.

Tropical rainforest occurs in Peninsular Malaysia (except in the northwest) and as small, often degraded forest remnants in Singapore. It has fluctuations in rainfall that may have a yearly pattern, but lacks an annual dry season that limits the growth of plants. In the zone below 800m, it takes the form of *lowland rainforest*. This is the tallest, most complex and most diverse forest of the region. Up to about 300m, this forest is commonly referred to as lowland dipterocarp forest because the dominant tall trees are mostly a mixture of species of the family Dipterocarpaceae. On the low ridges of hills above 300m it is commonly called hill dipterocarp forest.

From about 800–1,500m, the lowland rainforest gives way to *lower montane rainforest*. Trees in this zone are shorter, and epiphytic plants are abundant. At its lower elevations up to about 1,200m, it is commonly called upper dipterocarp forest, and at higher elevations, where dipterocarps are absent, it is called oak-laurel forest after the dominant tree species.

Beyond an elevation of 1,500m, there is often a rather marked transition to *upper montane rainforest*, characterised by short, small-leaved, flat-topped trees. This is also known as montane ericaceous forest, in which trees of the families Ericaceae (Heaths) and Myrtaceae (Myrtles) become abundant. Mosses also become more abundant, especially in sheltered areas and where cloud cover envelopes the forest frequently, giving rise to what is commonly called mossy forest or cloud forest, in which mosses dangle from trees and the ground is covered in a spongy layer of peat and dead sphagnum moss.

Tropical seasonal forests occur in the northwestern extreme of Peninsular Malaysia and throughout Thailand. In these forests, there is decreased rainfall and increased seasonality, with regular annual water stress of at least a few weeks. These forests can be broadly classified as deciduous, evergreen or mixed deciduous-evergreen. They are most easily distinguished during the dry seasons when they exhibit different degrees on leaf-fall.

Deciduous forests, also called monsoon forests, occur in the lowlands in northern Peninsular Thailand and parts of Continental Thailand, variably from about sea-level to 800m elevation. In these forests, the annual dry season results in a shortage of water that limits the growth of the plants and makes the forest largely deciduous. Beginning in December and particularly in March and April, the forest becomes nearly leafless. In its original undisturbed form, little of which remains, teak was often the dominant species in the deciduous forests. When heavily exploited for timber, the dominant canopy species change and bamboo becomes much more abundant in the understorey.

Mixed deciduous-evergreen forest occurs in the northwestern extreme of Peninsular Malaysia, most of Peninsular Thailand and parts of Continental Thailand. It has a mixture of evergreen and deciduous plants. Up to a third of the tall trees are deciduous at some time of the year. *Evergreen forest* occurs in Thailand at elevations above 1,000m. In comparison to other seasonal forests of Thailand, diversity is highest in this forest. Much of the evergreen forest has been cleared for traditional agriculture. The most pristine evergreen forests are in the upper mountain slopes and valleys through which streams pass. Forests growing along rivers have structure closer to a rainforest. When they occur among shorter seasonal forests, they are known as gallery forests. They tend to be evergreen and harbour species not found in the surrounding forest.

When the original forests are cleared, a succession of different secondary plant communities develop, resulting in a mosaic of other vegetation types. In Continental Thailand, where repeated clearing and burning has often taken place, it results in grasslands. Over an extended time, a scrub-like or low, open forest called *dipterocarp-oak forest* develops on these highly degraded sites, dominated by several species of dipterocarps with scattered oaks. As the forest becomes older, oak trees become more abundant.

Two other inland forest formations are found in the region. They are heath forests and forest occurring on limestone. **Heath forests** can be found along some of the coasts of S Thailand and Peninsular Malaysia, as well as on quartzite ridges, and also less commonly on plateaus and the gently sloping sides of some ridges. The trees are generally pole-sized and form a low canopy with an understorey that is often rich in epiphytes, small climbers, ant-plants, pitcher plants and mosses. **Forest on limestone** is usually easily recognised by the short, heath-like trees growing on steep, exposed and whitish rock faces and crags. Some plant species are specifically associated with the bases, slopes or summits of limestone hills.

Towards the coasts, specialised forest types occur. In some cases they may be narrow and fringed by inland forest types, especially in hilly coastal areas. In other areas, specialised coastal forest formations may extend for a great distance inland. **Beach vegetation** grows on the upper shores of beaches above the tide line. The plant species in this habitat are rarely found outside it. On the sandy shores, low creepers and herbaceous plants grow, behind which grow Casuarinas, coconut and low trees with wide spreading branches. **Mangroves** form at the estuaries of rivers where silt carried from upstream deposits at the river mouths, especially on the sheltered coasts of the west of the peninsula. The ground in mangroves is muddy and usually submerged at high tides. Trees tend to be relatively uniform in height, with special adaptations for the soft, unstable and waterlogged soils. At the inland edges of mangroves and the upper limits of the tides along river estuaries, *brackish-water forest* occurs. Mangroves are an important habitat for wildlife. However, most of the natural mangrove forests have been cleared for agriculture, aquaculture and settlement. What remain are usually small strips of mangrove fringe bounded by levies. Managed mangrove forests can be found in Matang in Peninsular Malaysia.

Peat swamp forests occur in the lowlands along upper river estuaries in E Sumatra, Peninsular Malaysia and Borneo. In Peninsular Malaysia, they are similar to rainforests and share many species in common, but are narrower in species. After logging, peat swamp forests become overrun with pioneer species and eventually reach a climax with a different composition of species. Vast areas of peat swamps in and around the region have been drained and cleared, especially for oil-palm cultivation. When drained, the dry peat soils become prone to spontaneous combustion during dry periods, resulting in deep peat fires that are difficult to control and blanket the region in haze. **Freshwater swamp forests** originally reached their greatest extent in Continental Thailand, especially along the Mekong River, but large areas of freshwater swamps have been cleared for paddy fields (rice) and other agricultural and plantation crops. Only small remnant patches remain. In the peninsula, the largest areas of freshwater swamps occur at Tasik Bera, Tasik Chini, and along the Sedili rivers in Johore. The vegetation in freshwater swamps varies greatly, but is more uniform in plant species composition than in dry-land forests.

Habitats

The different vegetation types in the region each support their own spectrum of butterfly species, though there is much overlap between them. Very little however, is known about actual habitat preferences of many butterfly species. In general the greatest numbers of species are found in the rainforests of the peninsula.

In the rainforests, montane elevations above about 1,000m have fewer butterfly species than lower elevations, but a number of species occur in these highlands that do not occur in the lowlands. Examples are the Cream Orange Tip, Green Commodore, Malayan

Commodore and Whitehead's Green Baron. In the lowlands, the Cream Orange Tip is replaced by the Yellow Orange Tip, which does not ascend to the highlands.

Also rich in butterfly species are the mixed deciduous-evergreen forests of Peninsular Thailand and NW Peninsular Malaysia. In Continental Thailand, evergreen forest occurring above an elevation of 1,000m tends to be richer in butterfly species than forest growing at lower elevations. This is because the deciduous forests of the lowlands are generally dominated in the canopy by a smaller range of tree species and have become degraded over the years by heavy exploitation.

Coastal wetland forests tend to be less rich in species. Many of the species that inhabit these habitats also flourish in parks, gardens, waysides and scrubland. However, wetland forests do support several species unique to these habitats, or that are rare in other habitats. An example is the Swamp Tiger, which is found only in or around mangroves.

Within a forest, butterflies may inhabit specific strata or habitats. Examples of each can be seen in the species descriptions that follow. However, they may sometimes fly higher or lower than usual. It is thought that species that fly in the canopies of trees may descend lower where there are small clearings, paths, rivers and open areas. Such gaps in the forest often encourage new growth and bushy plants that flower frequently, attracting butterflies to their nectar and to lay eggs on the fresh growth of the host plants. Consequently, territorial displays and sun-basking also take place in these areas.

Rivers and streams are also frequented by males of many species of butterflies that feed on salts in wet sand. Many butterfly species also fly up to hill- and mountain-tops to mate, while others may seek out sheltered gullies for this purpose. These behaviours are described in further detail below, but it is obvious that the habitat needs of butterflies are complex and that a range of habitats may be required for the survival of a single butterfly species.

Behaviour

The behaviour of butterflies often provides useful clues that can assist in their field identification. Most butterflies are active during daylight and prefer sunny weather, as they generally rely on the warmth of the sun to provide body heat sufficient for the vigorous activity of flight. Activity generally starts around 9 am in the morning and tapers off in the afternoon, becoming much lower after about 4 pm. However, some butterflies are more active in the late afternoon and early evening, or in the hours leading up to dusk. Often, different behaviour is observed at different hours. For example, feeding may take place twice in the day in the morning and afternoon while territorial displays may take place in the evening.

Some butterflies, particularly certain species of skippers and browns (Satyrinae), including owls (Amathusiini), are most active in flight at dawn and dusk and tend to be reclusive the rest of the day. This is referred to as a crepuscular habit. They may begin territorial flights before light dawns or at evening twilight and fly in somewhat fixed flight paths for an hour or so in the light of the morning or until complete darkness falls.

Different groups of butterflies have different flight patterns varying from slow gliding and gentle flapping to extremely fast wing-beats and wildly erratic flight paths. Skippers are well known for their fast, straight-line flight with rapid changes in direction. Many species of brushfoots in the admirals and relatives subfamily (Limenitidinae) have a flitting flight with intermittent wing flaps and wing glides. Brushfoots of the milkweed butterflies subfamily (Danainae), on the other hand, fly with leisurely wing flaps and seemingly effortless gliding.

When landing, different butterfly species may choose different locations and adopt different postures. They may land with wings closed, partially open, spread almost flat or even flexed slightly beyond horizontal. Skippers often rest with wings closed but flex the hindwings flat and forewings ajar when sun-basking. Most butterflies land on leaves, often on the upper surfaces, but sometimes underneath the leaves. Some are fond of landing on the trunks of trees, usually with head pointing downwards. Others prefer twigs and stems on which to rest. Invariably, when seeking shelter from rain, butterflies rest under leaves with their wings closed and will not move even if the leaves shake.

After rain, when the sun begins to shine, butterfly activity often spikes. The return of warm sunlight and the humid environment generated by rain makes it conducive for butterflies to fly and seek out flowers for nectar and damp ground for salts. After such weather, and in the first part of the morning, it is common to see butterflies sun-basking by opening their wings as they rest on a sunlit leaf. The veins that run through their wings carry blood that brings the heat of the sun back to their bodies. They may also occasionally bask on rocks, open ground and roads that absorb heat rapidly. When the weather becomes too hot, they can close their wings and orientate themselves in a manner that minimises sunlight on their wings. They also tend to fly into shadier areas and rest more.

A very obvious behaviour of some butterflies is what is called 'puddling'. Males land on the banks of rivers or on moist ground, especially where there is animal urine, to feed on water and salts. The salts are concentrated in the body while excess water is expelled in spurts from their abdomens. It is thought that this behaviour is important for reproductive processes in males, and that the nutrients gained are passed on to females in the sac that encloses the sperms. The phenomenon of puddling is made more spectacular by the fact that puddling butterflies tend to attract more puddling butterflies, which leads to large groups of closely packed butterflies congregating at particular spots, often in

Mass puddling of mainly whites and sulphurs (Pieridae) in Thailand.

mixed groups and a multiplicity of colours. Such group behaviour may regulate the physiology of individual butterflies and indirectly contribute to other phenomena like migrations.

Butterfly migrations involve the mass flights of individuals in a specific direction over a large distance, and sometimes even across seas. They may be diffuse and scattered and therefore difficult to detect, or funnelled into streams of butterflies and therefore very spectacular. They may be mixed in species or confined to a few closely related species. It is thought that migrations occur in response to growing populations and food plant scarcity in this region. Yearly seasonal migrations that occur in subtropical and temperate areas are rare in the region. As a consequence, migrations are infrequent, though they still play an important role in the health and dispersal of some butterfly populations. As forested areas decline, however, migrations have been observed even less frequently and on a less spectacular scale. The most commonly observed migrations are those of the Lemon Emigrant and Mottled Emigrant.

Most adult butterflies feed on nectar that they obtain from flowers. Nectar is rich in sugars that are easily converted to energy, and is also a source of amino acids for the building of body tissues and production of eggs. Larger butterflies tend to prefer larger tubular flowers, and smaller butterflies have to feed at small flowers unless they have a disproportionately long proboscis, such as that of the Chocolate Demon (p.146).

Many butterflies, however, are fond of feeding on the sap of injured tree trunks and the juice of rotting forest fruits, most often on the forest floor. In fact, there are some sap- and fruit-feeding species that are not known to feed on nectar. A number of butterfly species such as the brownwings (*Miletus*) and darkwings (*Allotinus*) actually feed on honeydew secreted by aphids and other plant sucking insects (pp.109–10). This honeydew is also fed on by ants that protect the insects, but which show little displeasure with the butterflies. These same butterfly species often feed on young plant-sucking insects when they are caterpillars.

The males of some species of butterflies are fond of feeding on animal excreta and carrion. These provide salts and nutrients in the form of amino acids. Skipper butterflies, especially the flats (Pyrginae), may seek out bird droppings on leaves or on the ground.

Male butterflies are often highly territorial. Some species form groups known as leks on one or more nearby trees and engage in spiralling aerial combats with each other. Other species patrol a larger territory singly, making back and forth or large circling flights along the banks of rivers and along forest paths. Territoriality in butterflies is an ecologically important behaviour that enables the fittest individuals to breed. It is most obvious on hill- and mountain-tops where many species can be seen flying back and forth from perches on low trees and passing over the summits. Most commonly, females pass over the summit while males may linger and defend perches. These sites also serve as a focal point for mating and enable many rare and sparsely distributed species to find mates and breed.

Mating is usually preceded by courtship rituals. These may involve males pursuing a female and hovering over it and may last for a long time. Sometimes, special scales, hairs or pouches on the wings and abdomens of the male are used to disperse scents used in courtship. In other butterflies, mating may occur soon after females emerge from their pupal cases, and the males seek out and await females even prior to their emergence. Most females can mate more than once. Others are plugged by males to prevent subsequent mating. In some cases, these plugs are removable by other males.

While males spend a lot of time engaging in territorial behaviour and seeking out mates, females spend the most time seeking out host plants and laying eggs. Butterflies are most often quite host specific as caterpillars. Carnivorous caterpillars also have specific insect prey. Adult female butterflies use visual and chemical stimuli to determine their hosts. A female that is searching for its host plant will usually move from plant to plant, touching different plants briefly as it flies. It may hover around a plant it thinks is a possible host plant, spending longer and landing briefly on it. Not only does it have to locate a suitable host species but also determine that there are suitable parts of the plant on which its progeny will feed. For example, the young caterpillars may require young leaves or shoots in the early stages of their development, or they may only feed on fruits or flowers. When the female finds and recognises its host plant, it will seek out suitable locations on which to lay its eggs. At this stage it may crawl on the leaf or stem and make brief and frequent flights. Its abdomen may be curled and will feel for a suitable place to deposit the egg. Once the egg is laid it will usually fly to another spot. However some species lay eggs in batches and tend to take longer to decide where to lay them.

DEFENCE AGAINST NATURAL ENEMIES

At every stage in their life cycle, butterflies are subject to threats from many natural enemies. They are preyed on by birds, lizards, spiders and insects such as praying mantises, dragonflies and robber flies. In addition, small wasps and flies that specialise in parasitising their eggs, larvae and pupae pose a constant threat. These insects are known as parasitoids because they eventually kill their host.

Butterflies have many means by which they protect themselves and escape from natural enemies, though not always successfully. Some groups of butterflies are poisonous to birds and lizards. They acquire and convert poisons from their host plants as caterpillars and sometimes also from nectar when they are adults. The milkweed butterflies are well known for their toxins, such as alkaloids, and cardiac glycosides, which act as heart poisons. The males even have special wing scales, and hair tufts that they can extrude from the end of their abdomen, which emit the powerful scent of such chemicals. These hair tufts are put into action when the butterfly is caught by a predator. Their bodies are resilient and can withstand light crushing by predators. They may even feign death. In the world of parasitoids, poisons are only of limited help in escaping attack. Virtually every poisonous butterfly species has equally specialised parasitoid species that can withstand the poisons or even make use of them. However, young predators that are susceptible to the poisons

A cerulean becomes prey to a spider.

Mimicry: *Top and bottom left:* Common Rose; *top right:* Great Windmill; *bottom right:* Common Mormon (female-form *polytes*).

learn quickly that these pungent butterflies are poisonous. To add to the effectiveness of this deterrent, these poisonous butterflies may have bright and contrasting colours, and they often resemble each other. They also fly very slowly as though advertising their colours. When colour is used to advertise distastefulness, it is referred to as warning or aposematic colouration. The avoidance of certain colour patterns by predators results in natural selection that enables evolution towards similarity in appearance. This similarity is known as Müllerian mimicry, after the German naturalist Fritz Müller who described the phenomenon. An example can be seen in the Great Windmill (p.24) and Common Rose (p.25), both of which are poisonous butterflies, and which resemble each other despite being in different genera. Butterflies that are perfectly edible to predators may mimic species that are poisonous, a type of mimicry called Batesian mimicry, named after the British naturalist Henry Bates who was the first to study the phenomenon in Brazil. These butterflies resemble and fly like the models that they mimic even though they are not poisonous. For this ruse to work, the numbers of individuals of such non-poisonous

Batesian mimics needs to be much smaller than the population of toxic Müllerian mimics, otherwise the deception breaks down. The female of the Common Mormon (p.28) has a form that is a Batesian mimic of the Common Rose. Mimes (p.26) and palmflies (pp.55–6) are other examples of Batesian mimics that mimic mainly milkweed butterflies.

Many butterflies rely on camouflage. The Blue Begum (p.75) is dappled shades of brown and dull green on the underside and blends with bark when it lands on a tree trunk. Other butterflies, such as the Common Duffer (p.71) may be brown on the underside with fine lines or numerous small, dark streaks that make them blend into a background of dead twigs and fallen leaves.

Some butterflies take resemblance to their environment to an extreme by resembling a whole leaf. This is seen most

A Blue Begum camouflaged against its favourite perch on a tree trunk.

impressively in the oakleafs and to a lesser extent the Autumn Leaf (pp.102–3). It also occurs quite commonly, though rather vaguely, in many species of browns (Satyrinae) and owls (Amathusiini) such as the Blue Catseye (p.63) and Saturn (p.69).

Some butterflies with cryptic undersides have medium to large sized 'eyespots' on their wings, for example the bush-browns, ringlets, rings, saturns and pansies, but perhaps most starkly in cyclopes (p.62), earning them their common name. These round markings are quite variable in appearance but are usually made up of concentric rings and a pale and dark inner area, the latter resembling the pupil of an eye. Large eyespots are thought to mislead predators into thinking they are the eyes of a larger animal. In addition, eyespots are used in recognising mates and selecting the best.

Eyespots in general, however, serve as decoys to detract predators from the vital parts of the butterfly. Predators such as birds, lizards and spiders aim for the head of a butterfly. When they mistakenly go for the eyespots, they end up with a mouthful of wing, which breaks off, enabling the butterfly to escape, albeit with less of its wings.

While many butterflies have wing markings and even wing shapes that give the impression of a head at the wrong end of the butterfly, false heads are taken to the most intricate extremes in blues (Lycaenidae), especially in the hairstreak subfamily (Theclinae). By flexing the hindwings when at rest, the butterflies are able to add to the deception by giving the appearance of a moving head. The tails often have a natural twist at their bases and therefore flex gently with the movement of the wings, giving

The lobes and tails at the end of a Club Silverline's tail produce a false head effect to detract predators.

the appearance of being able to move independent of the head. The false head of blues is very effective in deflecting attack away from their real head and other vital parts of their body.

Butterflies usually have more cryptic patterns on the undersides than on the uppersides of their wings. In fact, the upperside often has splashes of bright colours, especially in males. These are used for displays in courtship and mating. But like many characteristics of butterflies, such bright colour patterns can have other functions. They can startle or surprise a predator when the butterfly takes to flight or opens its wings briefly. When used for this purpose, they are referred to as deimatic patterns.

Flashes of bright colours also distract predators by focusing their attention on the colour, which then disappears when the butterfly closes its wings, confusing the predator. Used in this way it is known as flash colouration. In addition, broken patterns and splashes of colour combined with fast and erratic flight paths make it difficult for predators to follow and pursue their prey, especially in the low light conditions of the forest understorey.

LIFE HISTORY

Butterflies go through three very different life stages in their development to an adult: egg, larva and pupa. In their immature stages they bear no resemblance to an adult and are therefore said to undergo a complete metamorphosis at the pupal stage.

Female butterflies generally lay their eggs on or near the hosts of their young after mating and once their eggs are fully developed. This may be at the start of their life or after a period of sexual maturation. They may lay their eggs singly or in batches depending on the species. The egg is an embryo contained in a hardened shell. The shell may take many shapes and may be intricately sculptured with ridges and pits and may bear tiny projections, spines and hairs. Some butterflies coat their eggs with sticky and sometimes toxic glue to ward off predators.

The embryo within the egg develops into a larva, which is commonly called a caterpillar. It chews its way out when ready to hatch and consumes the egg shell for its

first meal. It then looks for its host or suitable parts of its host plant on which to feed. The most common hosts are plants but some are carnivorous in their early stages, feeding on plant-sucking aphids, plant hoppers, tree hoppers, mealy bugs or even ant larvae.
The caterpillar feeds on its host voraciously. It may do so at any time of the day but many species are nocturnal, feeding by night and resting by day and relying on camouflage (or aposematic colouration) to avoid being preyed upon. If the eggs are laid in a big batch the caterpillars may behave gregariously. They may feed communally, side by side, and walk in close contact with each other. As the caterpillar grows, it sheds and changes its skin (moults) a number of times. This is because its skin has a limited capacity to expand. The period between moults are called instars.

When it has developed to its full size, the caterpillar seeks out a location to enter a resting phase called the pupa or chrysalis. This may be under a leaf, in a rolled or folded leaf shelter, on a stem, in crevices on a tree trunk, on the ground under leaves or even in a shallow soil excavation, depending on the butterfly species. It anchors the end of its body to a surface using silk threads and, depending on the species, may also use silken threads to gird the middle section of the caterpillar and consequently pupa to the surface on both sides of its body. Thereafter, with a final moult, the caterpillar transforms into a pupa, which gradually takes a form, structure and colouration of its own as the outer shell of the pupa hardens.

Within the pupa, enzymes are released that digest most of the cells. Groups of cells called imaginal discs (after the word imago, which is the emerging adult), and that are unaffected by the digestive enzymes, multiply to form the organs and different body parts of the butterfly. When fully developed, the butterfly splits open the pupal casing and emerges. It excretes fluid waste matter left over from its developmental processes within the pupal case and pumps blood into the veins of its unexpanded wings. The wings expand as the veins expand and harden quite quickly. The newly emerged butterfly is then ready to make its maiden flight.

SEASONALITY

In Singapore and Peninsular Malaysia, the peak butterfly season is from April to July. In Thailand, peak season is between June to October. However butterflies can be seen throughout the year.

Seasonality relates back to life history, plant growth and annual variations in the climate. Even in the peninsula and Singapore where there is no marked wet and dry season, annual onsets of rain showers and storms from March to May cause growth spurts in plants. As the plants produce new leaves, butterflies benefit from increased food resources, since many butterflies grow best when feeding on the nutrient-rich and younger leaves or shoots of their host plants, which have less defensive chemicals than older leaves. As each new generation is produced, butterfly populations build up, often reaching a peak from May to July, but varying according to the species. In seasonal climates that have pronounced wet and dry months, plants are partially deciduous to a varying degree. This results in an even more pronounced seasonal effect with butterflies, which increase in numbers greatly after the dry season gives way to rains.

THE BUTTERFLY BODY AND WINGS

The body of an adult is divided into three parts: the head, thorax and abdomen. The head bears a pair of large compound eyes, antennae (or feelers) and the mouth parts. The thorax bears three pairs of legs (forelegs, midlegs and hindlegs) and two pairs of wings, while the abdomen holds the digestive, reproductive and many other important organs.

The antennae are used in sensory perception. The mouth parts consist of a long proboscis that can be curled up between two palpi. The proboscis functions as a tube to suck in fluids such as nectar, juice, sap and salt-enriched water. The legs of butterflies are jointed. At their tips they have small claws and pads to provide grip, and sensory organs that enable them to taste. The brushfoots (Nymphalidae) use only four legs for walking and standing. The small front legs are retracted against the thorax. This can be seen in the picture of the Pallid Faun (p.67), in contrast to the Great Windmill (p.24), which uses all six legs for standing.

The wings of butterflies are membranous with raised tubular veins and covered with numerous tiny scales that give colour and pattern to the wings. Some scales are coloured by pigments. Others have structural designs that enable refraction of light and produce iridescent colours. Wing areas with little or no scales show as translucent or transparent regions. Wing structure, colour and pattern are important in identification, although colour and pattern can also be variable within a species.

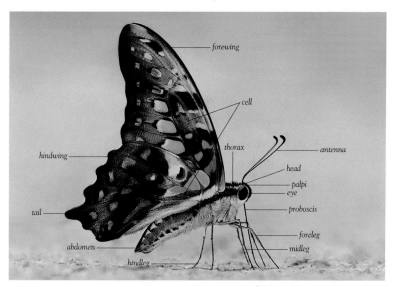

The parts of the body and wings of a butterfly.

The frontal and rear wings are referred to as the forewings and hindwings, respectively. The measurement given beside each species name is the forewing length, from its base to its tip (apex), and is given only as an indication of size. It is variable within a species but is simplified to an approximate average. Wingspan is generally estimated to be twice the forewing length plus the body width.

Different regions of the wing are referred to by different terms as shown in the illustration. Descriptions of the wing patterns frequently make use of these terms. For example, the upper margin of the wing is referred to as the costa, and a spot occurring on the costa may be referred to as a costal spot. The veins of the wings mostly radiate from two interconnected stems that enclose an area called the cell. In references on butterfly identification, the interspaces between the veins are identified by numbers (e.g., space 1b, space 2, etc.). In this field guide, however, the use of space numbers to describe the location of markings has been avoided because they can be difficult to make out in the field.

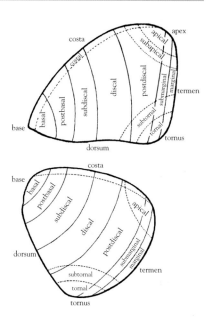

Regions of a butterfly's wings.

CLASSIFICATION AND NAMES OF BUTTERFLIES

Living things are classified into hierarchical groups that generally reflect their relatedness through a shared lineage or evolutionary pathway. The latter are inferred from key physical characteristics shared in common and, in recent years, genetic characteristics. For example, all the members of the order Lepidoptera, to which butterflies and moths belong, share in common the presence of scales on their wings and bodies. An order is conventionally divided into suborders, superfamilies, families, subfamilies, tribes, subtribes, genera, subgenera, species and subspecies, each being a component of the preceding group. The prefix 'infra' is sometimes used to denote a group beneath 'sub', for example infraorder, which is a grouping beneath suborders.

The order of butterflies in a checklist and book usually follows their classification and lineage, that is from the earliest evolutionary descent lines to the most recent. Because our knowledge of this is ever increasing, classifications and order are constantly being updated.

An up-to-date classification of genera and subgenera as we know it is given after the species descriptions in this book, but the arrangement of groups and species described is a compromise between this classification and familiar classifications used in the textbooks of the region. This is to provide readers with a degree of familiarity and similarity when cross referencing with these texts.

Butterflies comprise two superfamilies in the Lepidoptera, Papilionoidea (True Butterflies) and Hesperioidea (Skippers), and are sometimes taken to include the Hedyloidea (American Moth-Butterflies) of South America. In general, butterflies differ from moths in having antennae that expand or swell towards the tip. They also almost always have independently movable wings, whereas moths usually have hindwings that lack independent movement but rely on a bristle (frenulum) that connects the hindwing to the movable forewing. Most butterflies fly by day and most moths by night, but there are also day-flying moths and twilight-flying butterflies. An open-wing posture at rest is not a good field recognition character for moths.

The scientific names of all living things are based on Latin and Greek. The genus and species are the basic components of the name, for example, the name *Trogonoptera brookiana* for the Rajah Brooke's Birdwing implies the butterfly belongs to the genus *Trogonoptera*, and the species is *brookiana*. Added after the species name may be a subspecies name, for example *Trogonoptera brookiana albescens*. This is a reference to the Peninsular Malaysian subspecies called *albescens*. A subspecies (commonly abbreviated as 'ssp.', plural 'sspp.') is effectively a distinct race. They may differ in appearance but share enough fundamental characteristics to be the same species and may interbreed in a narrow range of their distribution, producing individuals intermediate in appearance. A subgenus name is sometimes used in parentheses after the genus name. The describer of the species (or subspecies when the subspecies name is used), is also sometimes added in non-italics after the name, with or without the year of description. Scientific names are a systematic means of identifying living things without ambiguity. They are more reliable than common names as they are regulated by the rules of the International Code of Zoological Nomenclature, in the case of animals. When changes are made to scientific names, it is intended to be for the improvement of classification and consistency in naming.

Common names though easier to remember and better accepted by nature enthusiasts, suffer problems of duplicate names for different species, multiple names for the same species and different names in different regions, even among regions that share a common language. Many new names have been created in recent years, sometimes increasing the problems. Common names may also not reflect relationship. For example, the Fuliginous Sailer *Lasippa monata* actually belongs to a group of butterflies called the lascars (*Lasippa* and *Pantoporia*) rather than to the group known as sailers (*Neptis*). The problems associated with common names have made some changes necessary in this book. The names generally follow Evans' (1927) *The Identification of Indian Butterflies*, and Corbet and Pendlebury's (1992) *The Butterflies of the Malay Peninsula* (4th edition). Other sources (listed in the Further Reading section) have also been used for species that were not named in these references. Names have been adapted or changed where they have been used twice for different species, have incorrect syntax, or are inappropriate or derogatory in our modern society.

BUTTERFLY WATCHING

Because of their small size, a pair of binoculars is very helpful for butterfly watching. When observing butterflies, take note not only of general colour and size but also of specific spot and band patterns on both sides of each wing, as far as possible. Flight behaviour as well as landing posture and location are also helpful in identification. Many butterfly enthusiasts replace a pair of binoculars with a camera. This has the advantage of giving a permanent record and a picture on which to base later identification when back from the field, providing the butterfly is sufficiently near to be photographed. Ideal cameras are single-lens reflex (SLR) with a macro lens and flash, but even compact cameras can sometimes produce a good enough photograph for a record. Stealth is usually needed to approach a butterfly to observe it closely. Abrupt movement scares butterflies away. Medium- to dark-coloured clothing is better than light, bright-coloured clothing for blending with the environment.

Well-planted gardens and parks provide many opportunities for seeing butterflies that may not be found in forests. The latter, of course offer the greatest diversity. Butterflies are usually readily seen from the edges of narrow forest paths and roads. Some even come into wider spaces such as clearings and wider roads if the edges have herbaceous flowering plants and low bushes. The edges of rivers and streams are also frequented by butterflies. At riverbanks, puddling butterflies make for a spectacular watch and provide great opportunities for photography. With patience, many species can be recorded just from the banks of rivers. Hill-tops also provide excellent opportunities for butterfly watching.

Baits can be used to attract butterflies that might otherwise be hard to see. Fruit is the most commonly used bait because they attract a wide range of species. Fruit bait needs to be overripe or rotting and sufficient in quantity to leave juice on the ground when crushed. A sprinkling of yeast or beer is sometimes added. It should be placed in a shady spot along a forest path. Prawn baits are used for species that visit carrion. Prawns rot very quickly and can be placed in a sunny spot on a riverbank or along a wide path. They attract butterflies for a narrow period of time after which they need to be replaced. All baits work best when they are used daily for several or more days, but will start to attract butterflies within hours or, in the case of rotting prawns, even within minutes if they are in the right location. Since baits lose their effectiveness with age, and wild animals like squirrels and monitor lizards consume them, longer periods of baiting require regular top-ups of the bait. Tying the bait in a piece of netting and hanging it on a tree helps reduce the problem of bait being eaten by animals.

WHERE TO GO

There are a great number of places in the region in which one can see butterflies. Only a selection can be mentioned here.

Peninsular Malaysia has a number of national and state parks. Taman Negara in the central part of the Peninsula is the largest and has a variety of inland forest types. Among the state parks are the Endau-Rompin State and National Parks, Selangor State Park, Royal Belum State Park, Perlis State Park and Penang National Park, which provide a good biogeographical coverage. The Krau Game Reserve in central Peninsular Malaysia can be visited at the Bukit Rengit and Kuala Lompat Wildlife Centres. Kuala Selangor Nature Park provides opportunities to see mangrove species, while the hill stations such as Fraser's Hill, Cameron Highlands, Maxwell Hill and forested areas of Genting Highlands are popular destinations for seeing both birds and butterflies that occur in montane habitats. Gunung (Mount) Brinchang in Cameron Highlands is one of the more accessible places in which upper montane forest species can be seen.

At the village of Ulu Geroh near Gopeng, spectacular congregations of the Rajah Brooke's Birdwing can be seen. It is also a good place to see other butterfly species. Local guides are compulsory. There are numerous recreational forests and forest reserves that can be visited, for example Panti Forest Reserve (including its Bird Sanctuary) in the southeast, Tasik Bera, Temengor, Kenyir, and the forest around The Gap. The islands off the coasts of the Peninsula, such as Langkawi, Tioman, Aur, Pemanggil, Perhentian and Redang also offer opportunities for seeing interesting butterflies.

For more information go to www.wildlife.gov.my and www.tourismmalaysia.gov.my.

Singapore has a few nature reserves, of which Bukit Timah Nature Reserve is central in the island. Many smaller inland parks such as Bishan Park, Dairy Farm Nature Park, Singapore Botanic Gardens, Lower Peirce Reservoir Park, MacRitchie Reservoir Park, Bukit Batok Nature Park and Fort Canning Park can be visited on the island. Mount Faber Park, Telok Blangah Hill Park and Kent Ridge Park are connected by the Southern Ridges, a series of trails and connecting bridges.

There are also a number of wetland and coastal reserves and parks in Singapore, for example, Sungei Buloh Wetland Reserve, Labrador Nature Reserve, Pasir Ris Park and Admiralty Park. The islands off Singapore, such as Pulau Ubin and Pulau Sentosa are also good sites for butterflies.

For more information go to www.nparks.gov.sg.

Thailand has very many National Parks, but not of all of them have good forest cover or support larger wildlife. Where they do, they are sometimes known as Wildlife Sanctuaries. A spectrum of well-forested national parks and wildlife sanctuaries are mentioned here. Hala-Bala Wildlife Sanctuary is in the far south in the Thai peninsula, but in the troubled Province of Narathiwat. Further north in central Peninsular Thailand is Khao Sok National Park. At the northernmost part of the peninsula is Kaeng Krachan National Park, the largest national park of Thailand, located at the borders of Burma in the eastern slopes of the Tenasserim Mountain Range. Large groups of butterflies puddle on its river banks. Erawan National Park is situated just north of the Thai peninsula in the Tenasserim Hills.

The Thung Yai – Huai Kha Khaeng Wildlife Sanctuary in the southern part of the Dawna Range is a complex of three wildlife sanctuaries that form what is called the Western Forest Complex. Located in the mid west of Continental Thailand, it has a variety of forest types and is a UNESCO World Heritage Site. Another World Heritage Site is Thailand's oldest national park, Khao Yai National Park, in central Thailand at the southwestern boundary of the Khorat Plateau. It forms part of the Dong Phayayen – Khao Yai Forest Complex, an area encompassing three national parks.

In southeastern Continental Thailand, Khao Soi Dao Wildlife Sanctuary is known for its large congregations of puddling butterflies, besides being an important refuge for other wildlife. This park is at the westernmost limits of the Cardamom Mountains that stretch into Cambodia. Just to the north of this park is Pang Sida National Park, which is also well known for butterflies. Even further north in East Thailand near the border with Cambodia is Phu Pha Yon National Park, which forms part of the Phu Phan mountain range.

North Thailand has a number of mountainous national parks, for example, Lam Nam Kok National Park in Chiang Rai, Thung Salaeng Luang National Park, Doi Khun Tan National Park, Mae Wa National Park and Doi Suthep – Doi Pui National Park. The latter is in the southernmost part of the Shan Hills in the foothills of the Himalayas.

For more information go to www.dnp.go.th/parkreserve.

OPPORTUNITIES FOR NATURALISTS

Although the taxonomy of butterflies has been relatively well studied compared to many other insect groups, it is likely that there are still species to be discovered in the region. In particular, many species complexes exist, where it is still uncertain whether different-looking individuals in a population are different species or variants of the same species. These problems may one day be solved by naturalists investigating life history and behaviour.

Many aspects of the natural history of butterflies are still unknown. Behaviour, such as nectaring, flight and resting position are usually inadequately described in the main reference materials on butterflies in the region, which tend to concentrate on taxonomy and species identification. In addition, relatively little is known about butterfly communities associated with different vegetation types, and niche preferences within forests. For example, the spectrum of butterflies that inhabit heath forests, peat swamps, mangroves, freshwater and brackish water swamps are barely known. Even the changes in populations that occur after different levels of degradation have still been little studied.

There is also a need for naturalists to develop field recognition characters for butterflies that will aid identification in the field. This is important not only for butterfly watchers but also for field researchers. Diagnostic characters of different butterfly species have usually been based on museum specimens. However, these characters are often not the most readily observable characters in the field.

Knowledge is only useful to others if it is shared. Technical information written up in a scientific format can be sent to journals for review and publication. Examples are the *Malayan Nature Journal*, *Serangga*, the *Raffles Bulletin of Zoology*, *Nature in Singapore*, the *Natural History Bulletin of the Siam Society*, *Tropical Natural History*, and the *Journal of Research on the Lepidoptera*. Some of these journals also publish electronically. Articles and photographs for the non-scientific community can be published in magazines such as the *Malaysian Naturalist*, *Conservation Malaysia* and *Nature Watch (Singapore)*. There are also a number of Internet forums that discuss butterflies, help with identification and host short articles. Examples are Butterfly Circle, www.butterflycircle.com/ and the Butterfly Interest Group of the Nature Society (Singapore), www.butterfly.nss.org.sg/home/. Many other outlets for information and articles are available and can be sourced online.

> **SWALLOWTAILS – PAPILIONIDAE**
> A well-known family of large- to medium-sized butterflies, sometimes with tails.

Rajah Brooke's Birdwing ▪ *Trogonoptera brookiana* 8.0cm

DESCRIPTION Unmistakable. Large, with elongate forewings. Male black with metallic green wing markings, touches of metallic blue, a bright red collar and red lower body streaks. The female of ssp. *trogon* resembles the male, but in ssp. *albescens* it is paler green

with a white apical forewing patch.
DISTRIBUTION Sumatra, Peninsular Malaysia and Borneo.
SUBSPECIES The ssp. *albescens* occurs in the western central foothills and mountain ranges from Perak to Negeri Sembilan in Peninsular Malaysia. Ssp. *mollumar* (sometimes considered to be the same as the Sumatran ssp. *trogon*) occurs in the east from SE Johor to SE Pahang and parts of Terengganu.
HABITS AND HABITAT Inhabits forest but flies into waysides. Males of *albescens* gather in spectacular groups on moist ground at certain rivers. Both sexes fly in the canopy, but may feed at low flowers.

Male of ssp. albescens

Common Birdwing
▪ *Troides helena* 8.0cm

DESCRIPTION Large. Forewing wholly black or with faint whitish apical streaks. Hindwing yellow with black veins and a wavy black border. All female *Troides* have large black spots inner to the black border.
DISTRIBUTION N India to S China, through Southeast Asia as far as Sulawesi.
SUBSPECIES The subspecies occurring in the region is *cerberus*.
HABITS AND HABITAT Inhabits lowland and montane forests, but comes to parks near forest. Flies slowly or swiftly, most often in the canopy, but comes low to feed at flowers.

Female

Golden Birdwing

■ *Troides aeacus* 7.0cm

Female of ssp. malaiianus

DESCRIPTION Distinguished from other similar birdwings by its smaller size, and the dark tornal dusting inner to the hindwing black border in the male. The **Malayan Birdwing** *Troides amphrysus* occurs in S Thailand and Peninsular Malaysia. It is larger with stronger whitish streaks on the forewing. The **Great Mountain Birdwing** *Troides cuneifera* from the highlands of Peninsular Malaysia is rare and has the thorax strongly reddened below.

DISTRIBUTION N India to S China and Taiwan, through Thailand, Malaysia and Indonesia.

SUBSPECIES The ssp. *malaiianus* occurs from Peninsular Malaysia to S Thailand and the larger ssp. *aeacus* occurs north of this range.

HABITS AND HABITAT Forested lowlands and foothills. Similar in behaviour to the Common Birdwing.

White-headed Batwing ■ *Atrophaneura sycorax* 7.5cm

DESCRIPTION A large and distinctive batwing with a prominent creamy white patch on the head and upper thorax. Abdomen creamy white; wings black, with outer half of the hindwing grey on the underside and grey-blue on the upperside, bearing large, black spots.

DISTRIBUTION S Burma, S Thailand, Peninsular Malaysia, Sumatra and W Java.

SUBSPECIES The ssp. *egertoni* occurs in this region.

HABITS AND HABITAT Rather rare, occurring in inland forests at low to moderate elevations. Like all batwings, it flies at a low to moderate height in the understorey, in clearings or at the forest edge.

Common Batwing ■ *Atrophaneura varuna* 5.5cm

DESCRIPTION Male bluish black with red markings on the underside of the body

and head. Female similar but larger, with whitish streaks on the forewing tornal area in ssp. *varuna*. The **Malayan Batwing** *Atrophaneura nox* from Peninsular Malaysia is similar but has rounder forewings, and is red only at the tip of the abdomen. The female has strong white streaks on the apex of the forewing.

DISTRIBUTION N India to Vietnam, Thailand and Peninsular Malaysia.

SUBSPECIES Occurs as ssp. *varuna* in Peninsular Malaysia and S Thailand, and as *zaleucus* in Central and N Thailand.

HABITS AND HABITAT Inland forests. Flies slowly or swiftly at low to moderate height in the forest understorey or at the forest edge. Sometimes active in the late evening.

Male of ssp. varuna

Male

Great Windmill
■ *Byasa dasarada* 6.5cm

DESCRIPTION Wings elongated; hindwing with wavy edges and a spatulate tail, a strong white spot and smaller, irregular tornal spots that are red towards the tail. Body and head red beneath. The very similar **Common Windmill** *Byasa polyeuctes*, which occurs in N Thailand where it is rarer than the Great Windmill, has all the small tornal spots red.

DISTRIBUTION Burma and N Thailand.

SUBSPECIES The Thai subspecies is *barata*.

HABITS AND HABITAT Rather slow flight with quick forewing beats. Males may settle on wet banks of streams. Both sexes visit flowers. Common where it occurs, inhabiting forests at all elevations. Usually seen in clearings and along streams.

Yellow-bodied Clubtail

■ *Losaria neptunus* 5.5cm

DESCRIPTION Easily recognised by its yellow
outer half of the abdomen that contrasts
against a black body. Forewing elongate, but
hindwing small, bearing a long club-shaped
tail with a thin stalk. Wings black with whitish
forewing patches and a small orange-red
hindwing patch.

DISTRIBUTION S Burma, S Thailand,
Peninsular Malaysia, Sumatra, Nias, Borneo
and Palawan.

SUBSPECIES Occurs as ssp. *neptunus* from
Peninsular Malaysia to southernmost
Thailand and as ssp. *manasukkiti* in the rest of
S Thailand.

Male

HABITS AND HABITAT Generally keeps to the understorey of heavy, inland forest in
the lowlands. Flies with a slow to swift, fluttering flight, pausing briefly when feeding or
searching for host plants. May come to the forest edge.

Common Rose ■ *Pachliopta aristolochiae* 5.0cm

DESCRIPTION Body red beneath. Wings black; hindwing with a central white patch and
pink-brown spots along its margin, which are larger and redder on the underside. Hindwing
with a moderately long, thin tail that expands slightly towards the tip. The **Common
Clubtail** *Losaria coon* (Thailand and Peninsular Malaysia) is larger and has a more elongate
forewing, a longer club-shaped tail and a larger, more broken, white hindwing patch.

DISTRIBUTION India and Sri
Lanka to South China, through
Thailand, Peninsular Malaysia,
Borneo, Sumatra, the Philippines
and the Lesser Sunda Islands.

SUBSPECIES The ssp. *goniopeltis*
occurs in most of Thailand, and
ssp. *asteris* occurs in S Thailand
and Peninsular Malaysia.

HABITS AND HABITAT Inland
forests and the forest edge in the
lowlands, sometimes venturing
into villages. Flies at low to
moderate height with a slow,
fluttering flight.

Ssp. asteris

Common Mime
■ *Papilio clytia* 4.5cm

DESCRIPTION Form *dissimilis* is glossy bluish black with greyish streaks and spots, and orange spots on the underside hindwing margin. Brown forms with streaks and spots reduced to the outer third of both wings or just the hindwing occur in Thailand and northwestern areas of Peninsular Malaysia. The smaller **Lesser Mime** *Papilio epycides* (N Thailand) has a browner underside than form *dissimilis* and only a single tornal orange spot.
DISTRIBUTION India to S China, Thailand to Singapore, the Philippines and Lesser Sunda Islands.
SUBSPECIES Occurs as ssp. *clytia*.
HABITS AND HABITAT Disturbed forests, parks and gardens. The larval food plant is a cinnamon. Flies slowly when undisturbed, mimicking the behaviour of milkweed butterflies.

Lime Butterfly
■ *Papilio demoleus* 4.5cm

DESCRIPTION Black-brown with yellowish-white spots and band. Hindwing underside with orange postdiscal spots edged with blue and black. Sexes similar.
DISTRIBUTION Eastern areas of the Arabian Peninsula to S China, Southeast Asia, Australia and Hawaii. Introduced to parts of the Caribbean and Central America.
SUBSPECIES The local subspecies is *malayanus*.
HABITS AND HABITAT Common in villages and may occur in gardens and parks in the lowlands wherever its larval host plant *Citrus* is grown.

Banded Swallowtail ■ *Papilio demolion* 5.0cm

DESCRIPTION Tailed. Black-brown with a green-tinted white band spanning both wings, breaking into rounded forewing spots. Hindwing margin with smaller white spots that may be orange-edged on the underside. Upperside with a small, orange tornal crescent or ring, which is more prominent in females.
DISTRIBUTION Burma, Indo-China, S China, Thailand, Sumatra, Peninsular Malaysia, Singapore, Borneo, Palawan, Java, Bali, Lombok and Australia.
SUBSPECIES The regional subspecies is *demolion*.
HABITS AND HABITAT Fairly common in undisturbed and disturbed forest up to the highlands. Flies fast, with a wavy and unpredictable flight path, at moderate height along the forest edge and in clearings.

Male

Black and White Helen

■ *Papilio nephelus* 6.0cm

DESCRIPTION Large, black, with a spatulate tail and prominent white hindwing patch. Lacks blue spots or scaling on the hindwing underside.
DISTRIBUTION N India to China, Thailand, Peninsular Malaysia, Sumatra, Java, Borneo and Palawan. Absent from Singapore.
SUBSPECIES Three regional subspecies, with intermediate forms: *sunatus* (Peninsular Malaysia south of the Kedah River), *annulus* (northwards to S Thailand), and *chaon* (the rest of Thailand). Ssp. *sunatus* has a white forewing band that is barely present in *annulus* and absent in *chaon*. All the underside markings are white in *sunatus*, while *chaon* has orange-yellow spots on the hindwing underside margin, and is often called the Yellow Helen.
HABITS AND HABITAT Inhabits forest, especially in the lowlands, flying fast at a moderate height with an erratic flight path. Sometimes seen on riverbanks or at puddles.

ABOVE RIGHT: *Ssp.* chaon
RIGHT: *Ssp.* sunatus

Red Helen ▪ *Papilio helenus* 6.5cm

DESCRIPTION Recognised by its full series of red spots on the underside hindwing margin. The **Blue Helen** *Papilio prexaspes* has blue scaling and orange spots beyond the white spots in ssp. *prexaspes* (S Thailand to Singapore). The montane **Lesser Helen** *Papilio iswaroides* (Peninsular Malaysia) has red spots confined to the hindwing tornus, and the large **Great Helen** *Papilio iswara* (S Thailand to Singapore) has blue spots above its tornal red spots.
DISTRIBUTION India and Sri Lanka to S Japan, and most of Southeast Asia. Absent from Singapore.
SUBSPECIES Represented by ssp. *helenus*.
HABITS AND HABITAT Common mainly in submontane and montane forest. Behaviour similar to the Black and White Helen.

Common Mormon ▪ *Papilio polytes* 4.5cm

DESCRIPTION Tailed. Male black with a white hindwing band. Female occurs in 2 forms in this region: female-form *cyrus* resembles the male but has a red tornal spot on the hindwing above, and female-form *polytes* has a white hindwing patch with a marginal ring of orange-pink spots, mimicking the Common Rose but lacking a red body. In Central Thailand, ssp. *pitmani* of the Blue Helen (*Papilio prexaspes*) resembles the male Common Mormon but lacks white spots on the forewing termen.

Male

DISTRIBUTION India and Sri Lanka to China, and throughout Southeast Asia.
SUBSPECIES A single ssp. *romulus* occurs in this region.
HABITS AND HABITAT Common in gardens, villages and disturbed forest in the lowlands. Its larval host plants include cultivated species of *Citrus*. Has a fast, restless flight, but female-form *polytes* flies slowly like the Common Rose.

Female-form polytes

Great Mormon
■ *Papilio memnon* 7.0cm

Male

DESCRIPTION Male tailless, black; upperside veins dusted with bluish-grey scales, especially on hindwing. Underside wing bases reddened, and tornus with a black-spotted red patch. Female polymorphic, the different forms mimicking different species of clubtails and batwings (*Losaria* and *Atrophaneura*). Female-form *butlerianus* is like the male but the forewing upperside has a wedge-shaped, red mark at the base, a whitish patch at the tornus, and white dusting along the veins. Female-form *esperi*, found mainly in the Peninsula and Singapore, is similar, but the whitish forewing patch is on the apical half of the wing instead of the tornus. A rarer and mainly Thai female-form, *agenor*, has elongated white spots on the hindwing. The tailed female-form *distantianus* is brown, and the hindwing has white spots and orange-pink edging.
DISTRIBUTION Sri Lanka and India to southern Japan, and throughout Southeast Asia.
SUBSPECIES Represented by ssp. *agenor*.
HABITS AND HABITAT Can be common in forest and at forest edges. Males fly fast and erratically at moderate height, often along upstream rivers, and may settle on riverbanks. Females fly more slowly mimicking the flight of their models.

TOP LEFT: *Male*. TOP RIGHT: *Female-form* butlerianus
ABOVE LEFT: *Female-form* esperi. ABOVE RIGHT: *Female-form* distantianus

Paris Peacock
■ *Papilio paris* 5.5cm

DESCRIPTION Tailed, with dark brown underside, metallic green dusted upperside and a large greenish-blue hindwing patch. Sexes similar. Two somewhat similar and rarer *Papilio* occurring in Thailand are the **Chinese Peacock** *Papilio bianor* (N and SE Thailand), which has an inwardly straight-edged blue patch that may also be inwardly diffuse, and the **Blue Peacock** *Papilio arcturus* (N Thailand), which has the blue patch diminished but very distinct, with 2 small pink spots below it.

DISTRIBUTION India to Thailand and S China. Also present in Sumatra, Borneo and Java but absent from Peninsular Malaysia and Singapore.

SUBSPECIES The subspecies occurring in Thailand is *paris*.

HABITS AND HABITAT Flies fast, but males come to wet sand at riverbanks, with wings quivering or spread open. Females visit flowers. Fairly common in forests up to the highlands, especially near streams and in open areas.

Banded Peacock ■ *Papilio palinurus* 4.5cm

DESCRIPTION Easily recognised. Tailed. Upperside black with a wide, metallic green band running down the forewing and hindwing, and with wings heavily dusted with metallic

green scales. Underside dark brown with black- and white-edged orange spots on the hindwing margin. Sexes similar.

DISTRIBUTION S Burma, S Thailand, Peninsular Malaysia, Sumatra, Borneo, Palawan and the Philippines.

SUBSPECIES Represented by ssp. *palinurus*.

HABITS AND HABITAT Primary and disturbed forest as well as secondary growth. Rare in Peninsular Malaysia, becoming commoner in the north. Flies very swiftly at a moderate height.

Kaiser-I-Hind ■ *Teinopalpus imperialis* 6.0cm

DESCRIPTION A very distinctive and striking butterfly. Sexes dissimilar. Mainly metallic green with a strong yellow-orange patch on the hindwing. The male has a prominent yellow-tipped green tail on the hindwing as well as 4 tooth-like projections, and has the underside forewing mainly orange-brown. The female has 2 prominent, pincer-like tails on the hindwing and has the underside forewing mainly grey.

DISTRIBUTION Nepal and NE India to N Thailand and SW China.

SUBSPECIES The subspecies occurring in Thailand is *imperatrix*.

Female of ssp. imperatrix *(Laos)*

HABITS AND HABITAT Fast-flying, keeping to the treetops of mountain peaks, but males may descend to the ground in the early morning. Generally rare but can be locally common. Inhabits montane forest above 1,500m, where its host plants, species of *Magnolia*, are found.

Glassy Bluebottle ■ *Graphium cloanthus* 4.5cm

DESCRIPTION Tailed. Black with very broad, transluscent blue-green bands on both wings, which break into large spots in the upper half of the forewing. Hindwing with a series of smaller, similarly-coloured marginal spots. Males and females are alike.

DISTRIBUTION N India eastwards through N Thailand, China and Taiwan. Also found in Sumatra. Absent from Peninsular Malaysia and Singapore.

SUBSPECIES The subspecies occurring in Thailand is *cloanthus*.

HABITS AND HABITAT Flies rapidly. Males may settle on wet ground, sometimes in groups, with wings closed. Females are sometimes seen feeding at flowers. Not widespread, but can be common where it occurs. Occurs in montane forests at elevations above 1,500m.

Male of ssp. cloanthus *(India)*

Common Bluebottle
■ *Graphium sarpedon* 4.0cm

DESCRIPTION Tailless. Black above, brown below, with a translucent, greenish-blue band on both wings, which is more continuous than in other tailless *Graphium*. Sexes similar.

DISTRIBUTION India to Japan, throughout Southeast Asia, the Papuan region and parts of Australia.

SUBSPECIES The local subspecies are *luctatius* (Singapore to S Thailand) and *corbeti* (the rest of Thailand).

HABITS AND HABITAT Common in forests, the forest edge and sometimes villages, at all elevations. Restless, flying fast with rapid wing-beats, even when feeding, but males may settle in numbers on riverbanks.

Common Jay ■ *Graphium doson* 4.0cm

DESCRIPTION Black-brown with translucent, greenish-blue bands and spots. Underside with red spots. Sexes similar. Distinguished from other similar blue-green *Graphium* by the fusion of the inner green hindwing streak with the adjacent central green band at their lower ends. The **Lesser Jay** *Graphium evemon* has the inner green streak separated from the central band and lacks a costal red spot, while the similar **Great Jay** *Graphium eurypylus* has a costal red spot. The **Striped Jay** *Graphium bathycles* has the central band bisected by the basal streak, and the very similar **Veined Jay** *Graphium chironides* has a slightly broader forewing band.

Ssp. evemonides

DISTRIBUTION Sri Lanka and India to Japan, through Thailand, Peninsular Malaysia, Singapore and most of Southeast Asia. Of the other species, only the Lesser Jay is found in Singapore.

SUBSPECIES Three subspecies: *evemonides* (S Thailand to Singapore), *axion* (in the north) and *kajanga* (Tioman Island).

HABITS AND HABITAT Similar to the Common Bluebottle in habitat and behaviour but keeps to forest.

Tailed Jay
■ *Graphium agamemnon* 4.5cm

DESCRIPTION Hindwing with a short tail.
Wings black above, brown below, with green
spots both above and below. Sexes similar.
The rather similar **Spotted Jay** *Graphium
arycles*, which occurs from the southern half of
Continental Thailand to Peninsular Malaysia,
lacks a tail and is slightly smaller.
DISTRIBUTION India to S China, throughout
Southeast Asia, the Papuan region and NE
Australia.
SUBSPECIES Occurs as the ssp. *agamemnon* from
Thailand to Singapore.
HABITS AND HABITAT Common at low
elevations in disturbed forest, the forest edge,
villages, parks and even in gardens. Larval host
plants include cultivated Custard Apple. Flight
fast, similar to the Common Bluebottle.

Fivebar Swordtail ■ *Graphium antiphates* 4.5cm

DESCRIPTION Easily recognised by its long sword-like tails and wing colours. Five other
species of swordtails occur in the region. The **Spectacle Swordtail** *Graphium mandarinus*
found in Thailand has a network of black markings on a white underside. Two others from
Thailand and Peninsular Malaysia are the **Fourbar Swordtail** *Graphium agetes*, with a
transparent patch bordered by black at the forewing apex, and the **Chain Swordtail** *Graphium
aristeus*, with heavy brown barring.

DISTRIBUTION Sri Lanka and
India to S China, through Thailand,
Peninsular Malaysia, Singapore,
Sumatra, Borneo, Palawan, Java and
the Lesser Sunda Islands.
SUBSPECIES Three subspecies:
itamputi (S Thailand to Singapore),
pompilius (to the north), and *pulauensis*
(islands of Tioman and Aur).
HABITS AND HABITAT Forest
and forest edges in the lowlands.
Flies swiftly at low to moderate
height, especially along forest tracks
and clearings. Males congregate on
wet ground.

Ssp. itamputi

Male

Malayan Zebra
■ *Graphium delessertii* 5.0cm

DESCRIPTION Black, with grey-white streaks punctuated by black spots. Hindwing tornal area with a small yellow patch. Female larger and greyer. Four other zebras, which are brown beneath, occur in the region. Males are more often seen. The male **Pendlebury's Zebra** *Graphium ramaceus* (S Thailand and Peninsular Malaysia) generally lacks grey streaks across the forewing cell, unlike other species. The **Great Zebra** *Graphium xenocles* (Thailand) has yellow near the hindwing tornus, which is absent in the **Lesser Zebra** *Graphium macareus* and in the smallest species, the **Spotted Zebra** *Graphium megarus* (both occur in NW Peninsular Malaysia and Thailand).
DISTRIBUTION S Thailand, Peninsular Malaysia, Sumatra and Borneo.
SUBSPECIES Occurs as ssp. *delessertii*.
HABITS AND HABITAT Inhabits forest up to the highlands. Males fly fast, sometimes settling in groups on riverbanks. Females fly slowly, mimicking the Smaller Wood Nymph.

Green Dragontail ■ *Lamproptera meges* 2.0cm

DESCRIPTION Small and elongate wings with very long hindwing tails. Black, with a transparent area in the outer half of the forewing, and a greenish-white stripe across both wings. Tail black, edged and tipped white. The **White Dragontail** *Lamproptera curius* (Thailand and Peninsular Malaysia) has whiter wing stripes and a narrower transparent area.
DISTRIBUTION Burma to S China, Thailand, Peninsular Malaysia, Sumatra, Java, Borneo,

the Philippines and Sulawesi.
SUBSPECIES The regional subspecies are *virescens* (Peninsular Malaysia and S Thailand) and *annamiticus* (Continental Thailand).
HABITS AND HABITAT Flies fast and low. Its rapid wing-beats, undulating flight and constantly quivering tails are unique. Inhabits forest at low to moderate elevations. Males may come to puddles.

> **WHITES AND SULPHURS – PIERIDAE**
> A family best known for its white, yellow and orange wing colours. Some are conspicuous garden butterflies.

Redbase Jezebel ■ *Delias pasithoe* 3.5cm

DESCRIPTION Black, with white upperside markings. Hindwing beneath with broad yellow markings and a red basal patch. The similar **Malayan Jezebel** *Delias ninus* from the highlands of Peninsular Malaysia and S Thailand has bluer upperside markings, and the red patch is also present above.

DISTRIBUTION N India to S China, through Thailand, Peninsular Malaysia, Singapore, Sumatra, Java, Borneo, Palawan and the Philippines.

SUBSPECIES Four regional subspecies: *parthenope* (Singapore to S Thailand), *beata* (Central Thailand), *pasithoe* (N Thailand) and *thyra* (E Thailand).

HABITS AND HABITAT Flies in the canopies of small trees. Commoner in the forested hills but can be locally and sporadically common in the lowlands. Sometimes found in villages and gardens.

Male of ssp. pasithoe

Redspot Jezebel ■ *Delias descombesi* 4.0cm

DESCRIPTION Easily recognised at rest or in flight by its strongly yellow hindwing underside marked with a conspicuous red costal streak. Male upperside white with diffuse dark borders. Female with forewing upperside dark and faintly white-streaked; hindwing white with a wide, dark border.

DISTRIBUTION NE India to Thailand and Peninsular Malaysia. Also occurs in the Lesser Sunda Islands.

SUBSPECIES Two subspecies in the region: *eranthos* (Peninsular Malaysia to S Thailand) and *descombesi* (further north).

HABITS AND HABITAT Highland forest. Most often seen flying in the canopy of small trees, but may come low to rest. Like all jezebels, it begins flight early in the morning.

Male of ssp. eranthos

Common Yellow Jezebel
■ *Delias baracasa* 3.0cm

DESCRIPTION Smaller than most jezebels. White, with underside of hindwing yellow, and veins dark-dusted. Female more heavily dark-dusted on the upperside. The rarer **Black and Yellow Jezebel** *Delias georgina*, which occurs in a number of subspecies on different mountain peaks in Peninsular Malaysia, is similar in size with rounder wings and dark blotches on the yellow hindwing underside, and the female has a much darker upperside. DISTRIBUTION Peninsular Malaysia, Sumatra and Borneo. SUBSPECIES Occurs as ssp. *dives* in Peninsular Malaysia. HABITS AND HABITAT Highland forest. Flies in the canopies of small trees. Can be common in hill stations such as Fraser's Hill in Peninsular Malaysia.

Painted Jezebel ■ *Delias hyparete* 4.0cm

DESCRIPTION White, with a yellow basal area and red marginal border on the hindwing underside. Veins dark-dusted. Female more heavily darkened on the upperside.
DISTRIBUTION India eastwards to S China and through Thailand, Peninsular Malaysia,

Singapore, Sumatra, Java, Borneo and the Philippines. SUBSPECIES The ssp. *metarete* is found from Singapore to Peninsular Thailand, north of which it is replaced by ssp. *indica* in which the yellow on the hindwing is more extensive. HABITS AND HABITAT Common in the lowlands, and occurs up to moderate elevations. Inhabits inland and coastal forest, including secondary growth, and is also found in villages, gardens and parks. The larvae live in groups and feed on mistletoes (species of *Dendropthoe*) that parasitise neglected garden and fruit trees.

Male of ssp. metarete

Psyche ■ *Leptosia nina* 2.0cm

DESCRIPTION Small, with rounded wings. Wing colour white, underside mottled with fine green-grey streaks, and upperside with a black wing tip and black spot. Sexes similar.
DISTRIBUTION Sri Lanka, India and Pakistan eastwards to S China, and through

Thailand, Peninsular Malaysia, Singapore, Sumatra, Java, the Lesser Sunda Islands, Borneo, the Philippines, Sulawesi and parts of Australia.
SUBSPECIES Occurs as ssp. *nina* from Thailand to NW Peninsular Malaysia and, southwards, as the less heavily marked ssp. *malayana*.
HABITS AND HABITAT Common in gardens, parks, villages, forest edges, forest roads and forest clearings from the lowlands to moderate elevations in the highlands. Flies low near the ground, with a slow, feeble flight that makes it easily recognisable.

Ssp. nina

Spotted Sawtooth ■ *Prioneris thestylis* 4.0cm

DESCRIPTION Forewing sharply pointed. Upperside of male black and white, extensively white on the hindwing and basal half of the forewing. Female with narrower white areas. Underside of both sexes with white areas reduced on forewing, and with broad yellow

streaks and spots on the hindwing.
DISTRIBUTION N India to S China, Continental Thailand and Peninsular Malaysia.
SUBSPECIES Occurs as ssp. *thestylis* in Thailand and *malaccana* in Peninsular Malaysia. Ssp. *thestylis* has less extensive white areas, and the female has a yellow tornal patch on the hindwing upperside.
HABITS AND HABITAT Occurs mainly in the highlands. Flies fast at low to moderate height. Males may be seen in small numbers among mixed groups of butterflies puddling at riverbanks and wet sand.

Male of ssp. malaccana

Male of ssp. themana

Redspot Sawtooth
■ *Prioneris philonome* 4.0cm

DESCRIPTION White with heavily blackened veins and a yellow basal patch on the hindwing underside. The extreme base of the hindwing underside also has a characteristic black-bordered red spot. Female is more heavily dark-dusted on the upperside.
DISTRIBUTION Sikkim to Thailand and Indo-China, Peninsular Malaysia, Sumatra, Java, Borneo and Palawan.
SUBSPECIES Occurs as ssp. *clemanthe* in Thailand and *themana* in Peninsular Malaysia, which has more extensively darkened veins and less extensive yellow on the hindwing.
HABITS AND HABITAT Fast-flying. Inhabits inland forest at all elevations. Most often seen along forest tracks and rivers, where males may settle on wet sand in mixed groups of other puddling whites and sulphurs.

Male of ssp. andersoni

Lesser Gull ■ *Cepora nadina* 3.0cm

DESCRIPTION Upperside of male white, with scalloped black borders; underside of hindwing deep orange with veins dark-dusted, forewing underside largely dark-dusted white. Female with veins heavily darkened on the upperside. The **Common Gull** *Cepora nerissa*, which occurs in NW Peninsular Malaysia and Thailand, is largely white below with the veins darkened yellow to brown.
DISTRIBUTION Sri Lanka and India to Thailand and Indo-China, Peninsular Malaysia and Sumatra.
SUBSPECIES Occurs as 2 subspecies in the region: *andersoni* in Peninsular Malaysia, in which the female has orange scaling along the hindwing dorsum above, and *nadina* in Thailand, which is very heavily darkened above.
HABITS AND HABITAT Inland forest, usually at moderate elevation. Flies moderately fast at low to moderate height. Settles on vegetation or at flowers, and males may come to puddles.

Orange Gull ■ *Cepora iudith* 3.0cm

Male of ssp. malaya

DESCRIPTION Similar to the Lesser Gull, but has a bright orange-yellow patch along the tornus and dorsum on the hindwing upperside, a brighter orange hindwing underside and a chocolate-brown border on the underside termen of both wings.
DISTRIBUTION S Burma, Thailand, Peninsular Malaysia and throughout the Southeast Asian islands.
SUBSPECIES Occurs as ssp. *malaya* in Peninsular

Male of ssp. malaya

Malaysia and S Thailand, and as ssp. *lea* in the rest of Thailand. On the islands of Tioman and Aur off the east coast of the Peninsula, it occurs as sspp. *talboti* and *siamamensis*, respectively, in which the females are heavily dark-dusted, especially in the latter.
HABITS AND HABITAT Similar to the Lesser Gull but restricted to the lowlands.

Forest White ■ *Phrissura cynis* 3.0cm

Male of ssp. cynis

DESCRIPTION Upperside of male white with a black border at tip of forewing. Underside similar but border paler, and often with a variable greyish to greenish basal scaling, and a grey-green discal line. Female brown with variable white markings that include a white forewing bar and brown-dusted white basal areas.
DISTRIBUTION S Thailand, Peninsular Malaysia, Sumatra and Borneo.

Male of ssp. cynis

SUBSPECIES The ssp. *cynis* is found in southernmost Thailand and Peninsular Malaysia, but on the islands of Tioman and Aur it occurs as the darker ssp. *pryeri*.
HABITS AND HABITAT Can be locally abundant in some lowland forests where it generally flies low near the ground, flying higher to feed at flowers. Males come to wet ground on forest paths on sunny days after a rainy night.

Chocolate Albatross
■ *Appias lyncida* 3.0cm

DESCRIPTION Male with hindwing underside yellow and forewing underside white, with broad chocolate brown borders on both wings. Upperside white with termen edged black. Female with yellow underside paler; upperside brown with white streaks, varying in extent.

DISTRIBUTION Sri Lanka and India to S China, through Thailand, Peninsular Malaysia and the Southeast Asian islands.

SUBSPECIES From Singapore to S Thailand it occurs as sssp. *vasava*, and in the rest of Thailand, as ssp. *eleonora*.

Male of ssp. vasava

HABITS AND HABITAT Common in inland forest at all elevations, including disturbed forest.

Striped Albatross
■ *Appias libythea* 3.0cm

DESCRIPTION Male white, with the outer ends of the veins blackened on the upperside, and virtually all the veins blackened on the underside. Hindwing underside with a small yellowish streak at the base. Female brown with white markings, paler and dusted with yellow scaling over the underside.

DISTRIBUTION Sri Lanka and India to Thailand, Peninsular Malaysia and Singapore.

SUBSPECIES Occurs as ssp. *olferna* (sometimes considered a species).

HABITS AND HABITAT A common butterfly found in gardens, roadsides, villages, mangrove edges, secondary forest and the forest edge at low elevations.

ABOVE LEFT: *Male*
LEFT: *Female*

Orange Albatross ■ *Appias nero* 3.5cm

DESCRIPTION Easily recognised. Dark orange to brick red above. Paler reddish orange beneath. The male is largely unmarked, but has dark scaling along the veins on the forewing and along the margin of the hindwing. The female has, on the upperside, black borders on both wings and black markings on the forewing.
DISTRIBUTION NE India to Burma, Thailand, Peninsular Malaysia, Singapore, Sumatra, Java, Borneo, the Philippines, Sulawesi and Buru.
SUBSPECIES Two subspecies: *figulina* in Peninsular Malaysia and S and SE Thailand, and the paler *galba* (sometimes considered a species) to the north.
HABITS AND HABITAT Fairly common in inland forest at all elevations. May come to the forest edge. Often seen in small numbers in groups of puddling whites and sulphurs.

Male of ssp. figulina

Malaysian Albatross ■ *Saletara liberia* 3.0cm

DESCRIPTION Male forewing pointed. Underside creamy lemon white to yellow, upperside white with a narrow black margin from the costa to tornus. End of abdomen with a pair of black hair tufts that can be seen as black streaks against the pale abdomen when the wings are open. Female with broad black borders on the upperside, and the hindwing may be yellow.
DISTRIBUTION Nicobars, S Thailand, Peninsular Malaysia, throughout the Indonesian Archipelago and in New Guinea.
SUBSPECIES The ssp. *distanti* occurs in this part of its range.
HABITS AND HABITAT The species may be locally common in inland forest at low elevations, and males may congregate in numbers on wet sand at the edges of rivers.

Male

Male of ssp. verna

Males of ssp. verna

TOP AND ABOVE: *Male of ssp.* aturia

Yellow Orange Tip

▪ *Ixias pyrene* 3.0cm

DESCRIPTION Underside yellow, mottled with brown; upperside yellow in ssp. *birdi* and variable yellowish white in *verna*, with broad black borders and an orange forewing bar (yellow in females of *birdi*). Females have narrower bars. The **Cream Orange Tip** *Ixias alticola* (highlands of Peninsular Malaysia) is similar in pattern but is white above, with the bar orange in the male and white in the female.

DISTRIBUTION Sri Lanka and India to S China, through Thailand, Peninsular Malaysia, Sumatra, Java, Borneo and the Philippines.

SUBSPECIES Two regional subspecies: *verna* (Thailand and NW Peninsular Malaysia) and *birdi* (the rest of Peninsular Malaysia).

HABITS AND HABITAT Local in occurrence at low altitudes in inland forest. Flies fast but low, settling on flowers to feed. Males come to puddles.

Great Orange Tip

▪ *Hebomoia glaucippe* 5.0cm

DESCRIPTION Large, white with a black-bordered orange band across the forewing apex on the upperside. Underside with fine dark lines. Female with dark spots on the upperside inner to a dark border. Island subspecies are yellow and may have a whitish band in the female.

DISTRIBUTION Sri Lanka and India to S China, through Thailand, Peninsular Malaysia, Borneo, the Indonesian Archipelago and the Philippines. Very rare in Singapore.

SUBSPECIES Four subspecies in the region: *glaucippe* (Continental Thailand), *aturia* (S Thailand and Peninsular Malaysia), *anomala* (islands of Tioman and Aur) and *theia* (Pemanggil Island).

HABITS AND HABITAT Uncommon. Fast on the wing, flying at moderate height. Inhabits coastal and inland dipterocarp forest up to moderate elevations. Males visit damp spots along riverbanks.

Malayan Wanderer

▪ *Pareronia valeria* 4.0cm

DESCRIPTION Males and females differ greatly. Upperside of male pale greenish blue with a black border and blackened veins; underside whiter and barely marked. Female upperside brown with grey-white streaks and yellow wing bases, mimicking the Yellow Glassy Tiger. The **Common Wanderer** *Pareronia hippia* (formerly called *anais*), which occurs in Continental Thailand and NW and NE Peninsular Malaysia, has the veins more thickly darkened in the male, and the female occurs in 2 forms that has or lacks yellow on the hindwing.

DISTRIBUTION S Burma, S Thailand, Peninsular Malaysia, Sumatra, Java, Borneo and Palawan.

SUBSPECIES Occurs as a single ssp. *lutescens* in this region.

HABITS AND HABITAT Males fly very fast along tracks in forest at low to moderate elevations. They may settle on wet ground. Females are rarer and mimic the slow, casual flight of the Yellow Glassy Tiger unless disturbed.

TOP: *Male.* ABOVE: *Female*

Mottled Emigrant ▪ *Catopsilia pyranthe* 3.5cm

DESCRIPTION Upperside greenish white with a forewing cell-end spot and narrow black border that is widest at the forewing apex and wider in the female. Underside a darker shade of greenish white, heavily mottled with fine brown streaks.

DISTRIBUTION Sri Lanka and India eastwards to S China and southwards through Thailand, Peninsular Malaysia, the Southeast Asian islands and NE Australia.

SUBSPECIES A single ssp. *pyranthe* occurs in the region.

HABITS AND HABITAT Gardens, parks, villages, scrubland and secondary forest at low elevations. Flies fast at moderate height. Migratory. The host plants are species of *Senna*, especially Seven Golden Candlesticks *Senna alata*.

Male

Male-form alcmeone

Lemon Emigrant
▪ *Catopsilia pomona* 3.5cm

DESCRIPTION Occurs in a number of distinct forms that are pale yellowish green to orange-yellow, with black, upperside borders of variable extent. Reddish-brown rings or patches are present on the underside of *pomona* forms but are absent in *crocale* forms.

DISTRIBUTION India to S China, through Southeast Asia as far as the Solomon Islands and northern Australia. Also occurs in Madagascar and Mauritius as a distinct subspecies.

SUBSPECIES The regional subspecies is *pomona*.

HABITS AND HABITAT Fast-flying at moderate height. Migratory. Inhabits open country, occurring in gardens, villages, parks, waysides and the forest edge. Often seen where its ornamental host plants *Senna* are grown.

Male

Orange Emigrant
▪ *Catopsilia scylla* 3.0cm

DESCRIPTION Upperside forewing white with a black border, hindwing bright orange. Underside deep orange with diffuse, dark spots and rings.

DISTRIBUTION S Burma, Thailand, Peninsular Malaysia, Singapore, Borneo, the Indonesian Archipelago, the Philippines and northern Australia.

SUBSPECIES A single ssp. *cornelia* occurs in the region.

HABITS AND HABITAT Similar to the Lemon Emigrant.

> ### Grass Yellows – Eurema
> Small yellow butterflies with irregular black forewing borders and narrow, black hindwing borders on the upperside, and small dark spots on the underside. Females paler, with borders inwardly diffuse. Eleven rather similar-looking species occur in the region. Slow- to moderately fast-flying at low height. May congregate on moist ground.

Common Grass Yellow
▪ *Eurema hecabe* 2.2cm

DESCRIPTION Underside without a chocolate forewing patch and with two small forewing cell spots (may be faint or worn). The **Three-spot Grass Yellow** *Eurema blanda* has narrower black borders and 3 cell spots.
DISTRIBUTION Central Africa to Japan, through Southeast Asia and New Guinea as far as N and E Australia, Fiji and Tonga.
SUBSPECIES The numerous Asian mainland subspecies described are now considered to belong to a single variable ssp. *hecabe*.
HABITS AND HABITAT Inhabits almost every habitat at all elevations.

Male

Chocolate Grass Yellow
▪ *Eurema sari* 2.1cm

DESCRIPTION Similar to the Common Grass Yellow but with a clear, undivided chocolate patch on the underside forewing apex. The **Changeable Grass Yellow** *Eurema simulatrix* has a clear but divided chocolate patch, while the **One-spot Grass Yellow** *Eurema andersonii* and the smallest species, **Talbot's Grass Yellow** *Eurema ada*, are among a few species that have just a small chocolate streak. All 3 species occur from Singapore to Thailand.
DISTRIBUTION Sri Lanka and India to Indo-China, Thailand, Peninsular Malaysia, Singapore, Sumatra, Java, Borneo and Palawan.
SUBSPECIES Occurs as ssp. *sodalis*.
HABITS AND HABITAT Inhabits inland forest at low to moderate elevations.

Male

Male

Malayan Grass Yellow
■ *Eurema nicevillei* 2.4cm

DESCRIPTION Easily recognised in flight as it is the only species in the region with a black upperside forewing border that extends along the dorsum. Previously placed as a ssp. of *Eurema tilaha*, which is now considered a different species restricted to Java and Bali.
DISTRIBUTION S Burma and S Thailand to Peninsular Malaysia, Sumatra, Borneo and the Mentawi Islands.
SUBSPECIES Occurs as ssp. *nicevillei* in this region.
HABITS AND HABITAT Stronger on the wing than most grass yellows, and inhabits lowland forest.

Tree Yellow ■ *Gandaca harina* 2.3cm

DESCRIPTION Very pale yellow above and below. Unlike grass yellows, the black border of the Tree Yellow is confined to the apex on the forewing upperside and is generally

Male of ssp. distanti

absent on the hindwing. The forewing border is regular in the male but has a tooth-like projection in the paler female. The underside is unmarked.
DISTRIBUTION NE India to Indo-China, through Thailand, Peninsular Malaysia, Singapore, Borneo, the Indonesian Islands and New Guinea.
SUBSPECIES There are 2 mainland subspecies in the region: *distanti* in Peninsular Malaysia and S Thailand, and *burmana* north of this. On the islands of Tioman and Aur off the east coast of the Peninsula occurs a further ssp. *aora*.
HABITS AND HABITAT Higher- and faster-flying than grass yellows. Occurs from the lowlands to mid-elevations in inland forest. Like grass yellows, males may settle on moist ground or wet sand.

BRUSHFOOTS – NYMPHALIDAE
The largest and most varied family of butterflies. They derive their name from their reduced front legs that are not used when walking.

MILKWEED BUTTERFLIES – DANAINAE
A subfamily of poisonous butterflies with warning colours. They resemble each other to enable easy recognition by predators and are mimicked by some non-poisonous butterflies.

Plain Tiger ■ *Danaus chrysippus* 4.0cm

DESCRIPTION Orange, with the wing margins and apical third of the forewing above black, spotted with white, and with a white forewing band. Male with a raised black hindwing spot below the 3 black discal spots. Form *alcippoides*, which has a largely white hindwing, is found in S Thailand and Peninsular Malaysia south of Penang, while the all-orange form occurs in Continental Thailand, NW Peninsular Malayisa and Singapore.

DISTRIBUTION Widely distributed from Greece and tropical Africa eastwards to China, through tropical Asia to the Australian region.

SUBSPECIES Occurs as ssp. *chrysippus*.

HABITS AND HABITAT Slow-flying. Locally common in villages and coastal scrubland especially where Crown Flower *Calotropis gigantea*, a favourite host plant, is grown.

Form alcippoides, *male*

White Tiger ■ *Danaus melanippus* 4.0cm

DESCRIPTION Black, with white marginal spots, a white forewing band, an orange-streaked forewing and white-streaked hindwing. Male with a raised, black hindwing spot. The **Common Tiger** *Danaus genutia* has an orange-streaked hindwing and narrower black wing borders.

DISTRIBUTION Sri Lanka and India to China and Southeast Asia up to Sulawesi.

SUBSPECIES The local race is *hegesippus*.

HABITS AND HABITAT Common in secondary and disturbed forest up to submontane elevations. Flies very slowly in open areas.

Male

Swamp Tiger
■ *Danaus affinis* 3.5cm

DESCRIPTION Resembles and is easily mistaken for the White Tiger (p.47) but can be recognised by its smaller size, band-like white hindwing patch and wedge-shaped orange-brown spots on the underside hindwing, which make the termen appear more orange-brown in flight.

DISTRIBUTION S Thailand to S Vietnam, Peninsular Malaysia, the Southeast Asian islands, Australia and New Caledonia. Absent from Singapore.

SUBSPECIES The race occurring in Thailand and Peninsular Malaysia is *malayanus*.

HABITS AND HABITAT Restricted to specific mangroves where it occurs in open areas and at the forest edge, flying slowly at low height. Breeds and nectars only on specific mangrove-inhabiting plant species.

ABOVE LEFT AND LEFT: *Female*

Dark Blue Tiger ■ *Tirumala septentrionis* 5.0cm

DESCRIPTION One of 3 species of *Tirumala* in the region which can be recognised by their short, thin V-shaped central markings on the hindwing. Black above and brown below with light blue translucent streaks and spots. Males have a darkened patch within the V-shaped markings.

DISTRIBUTION Afghanistan to India and Sri Lanka across to S China and Taiwan, through Thailand, Peninsular Malaysia, Sumatra, Java, Borneo and Palawan. Absent from Singapore.

SUBSPECIES Occurs as the nominate ssp. *septentrionis* in this region.

HABITS AND HABITAT Slow-flying, usually keeping low. The least rare species of *Tirumala*, occurring in forested areas at all elevations. Commoner in Thailand and the north of Peninsular Malaysia, but sporadically and locally abundant in some areas of the Peninsula.

Male

Yellow Glassy Tiger ■ *Parantica aspasia* 4.0cm

DESCRIPTION Black with broad, translucent streaks and spots that are yellow on the hindwing and the base of the forewing, and grey-white towards the tips. Males have more extensive yellow, more elongate forewings and a double, dark tornal spot on the hindwing. Very distinctive, but the less often seen female Malayan Wanderer (p.43) is a good

mimic of this species. The **Glassy Tiger** *Parantica aglea*, in which the translucent markings are completely pale bluish white, is common in Thailand and NW Peninsular Malaysia.

DISTRIBUTION S Burma, Thailand and southern Indo-China, Peninsular Malaysia, Singapore, Sumatra, Java, Borneo and Palawan. Very rare in Singapore.

SUBSPECIES Represented by ssp. *aspasia*.

HABITS AND HABITAT Common where it occurs in lowland and montane forest. Flies very slowly at low to moderate height.

Female

Dark Glassy Tiger ■ *Parantica agleoides* 4.0cm

DESCRIPTION Black-brown with slightly translucent, thin, grey-white stripes and spots. Lacks a bluish tinge in flight. Differentiated from similar-looking species of *Ideopsis* by the

unbroken greyish streaks in the forewing cell, which are characteristic of *Parantica*, and from other species of *Parantica* by the narrower wing streaks. Males have a double, black tornal ring on the hindwing underside.

DISTRIBUTION S Burma, S Thailand and southern Indo-China, Peninsular Malaysia, Singapore, Sumatra, Java, Borneo and Palawan.

SUBSPECIES Occurs as ssp. *agleoides*.

HABITS AND HABITAT Common in disturbed and primary coastal and inland forest up to moderately high elevations. Flies slowly in the understorey or forest edge at low to moderate height.

Female

Chocolate Tiger ■ *Parantica melaneus* 5.0cm

DESCRIPTION Upperside black with broad, translucent wing streaks and spots. Underside similar but paler, with the black areas of the wing margins and forewing apex broadly replaced with chocolate brown. Larger than the preceding species of *Parantica*. Males have a small, black tornal patch. Mimicked by the rarer Lesser Mime (p.26) in Thailand.

DISTRIBUTION N India eastwards to Indo-China, through Thailand, Peninsular Malaysia, Sumatra, Java, Borneo and Palawan. Absent from Singapore.

SUBSPECIES The race *sinopion* occurs in Peninsular Malaysia, and *plataniston* in Thailand.

HABITS AND HABITAT Common in or near forested areas at moderate to high elevations. Frequently seen feeding at flowers along roadsides or in gardens in hill stations. Flies slowly at low to moderate height.

Male

Chestnut Tiger ■ *Parantica sita* 5.0cm

DESCRIPTION Resembles the Chocolate Tiger in wing markings and size, but the upperside of the hindwing has a wide, chestnut-brown band along its outer margin. Males have a black patch near the hindwing tornus. On mountain tops and ridges it may be confused with its rarer mimic, the Tawny Mime (p.26).

DISTRIBUTION N India to SE China and Japan, through Thailand and Peninsular Malaysia. Absent from Singapore.

SUBSPECIES Three subspecies in the region: *melanosticta* in most of Continental Thailand, *sita* in N Thailand, and *ethologa* in S Thailand and Peninsular Malaysia.

HABITS AND HABITAT Restricted to highland forest. Commonly seen visiting flowers in gardens and along roadsides in forested hill stations, alongside the more abundant Chocolate Tiger.

Male of ssp. sita

Blue Glassy Tiger
■ *Ideopsis vulgaris* 4.0cm

DESCRIPTION Resembles the Dark Glassy Tiger (p.49), but appears slightly bluish in flight, and has a broken forewing cell-stripe. Differs from the bluer **Ceylon Blue Glassy Tiger** *Ideopsis similis* (Central Peninsular Malaysia to southern Continental Thailand) in its narrower forewing discal spots. The **Grey Glassy Tiger** *Ideopsis juventa* (east coast and eastern islands of Peninsular Malaysia) lacks a blue tinge and, unlike other *Ideopsis*, its hindwing cell-stripe is barely if at all split.
DISTRIBUTION S Burma to Indo-China, through Thailand, Peninsular Malaysia, Singapore, Sumatra, Borneo, Palawan, Java and the Lesser Sunda Islands.
SUBSPECIES Occurs as ssp. *macrina* from Singapore to S Thailand, and as the Indo-Chinese ssp. *contigua* in parts of E and SE Thailand.
HABITS AND HABITAT Flies slowly in open areas, inhabiting disturbed and primary coastal and inland forest up to moderately high elevations.

TOP AND ABOVE: *Ssp*. macrina

Smaller Wood-Nymph ■ *Ideopsis gaura* 5.0cm

DESCRIPTION Wings translucent, greyish white with rounded black spots. Much smaller than similarly marked tree nymphs (p.52). Females are paler and have more rounded wings.
DISTRIBUTION S Thailand, Peninsular Malaysia, Sumatra, Java, Borneo and the Philippines. Thought to be extinct in Singapore.
SUBSPECIES Represented by ssp. *perakana* in S Thailand and Peninsular Malaysia, and by ssp. *kajangensis* on the island of Tioman off the eastern coast of the Peninsula.
HABITS AND HABITAT Glides slowly at a moderate height, with slow, intermittent wing flaps. Inhabits inland forest from the lowlands to highlands.

Female of ssp. perakana

Ashy-white Tree-Nymph ■ *Idea stolli* 8.0cm

DESCRIPTION Wings large; translucent greyish white, with black spots. Females with rounder wings. The **Tree-Nymph** *Idea lynceus* is a darker, smoky grey colour, while the **Malayan Tree-Nymph** *Idea hypermnestra* is whiter with rounder wings. Both occur in Peninsular Malaysia and S Thailand. The **Large Tree-Nymph** *Idea leuconoe*, with

interconnected black marginal spots, inhabits some mangroves in Singapore, Tioman Island and Central Thailand.
DISTRIBUTION Extreme south of Thailand, Peninsular Malaysia, Singapore, Sumatra, Java, Borneo and Palawan.
SUBSPECIES The single regional subspecies is *logani*.
HABITS AND HABITAT Flies with slow wing-beats, often gliding very slowly, usually in the canopy of trees, but may descend lower. Inhabits forest at low to moderate elevations.

Male

CROWS – EUPLOEA
A genus of milkweed butterflies known for their dark and often blue-shot wing colours. Fifteen species occur in Thailand and Peninsular Malaysia. Half also occur in Singapore.

Spotted Black Crow
■ *Euploea crameri* 5.0cm

DESCRIPTION Brown with white, oval, forewing spots and white dots on the hindwing margin.
DISTRIBUTION S Burma, S Thailand, southern Indo-China, Peninsular Malaysia, Singapore, Sumatra, Java, Borneo and Palawan.
SUBSPECIES Two subspecies: *bremeri* (S Thailand to Singapore) and *crameri* (islands of Tioman and Aur).
HABITS AND HABITAT Inhabits coastal and inland forest at low elevations. Flies slowly. Locally common, for example on the east coast of Peninsular Malaysia and in Singapore.

Ssp. bremeri

Lesser Striped Black Crow
■ *Euploea eyndhovii* 4.5cm

DESCRIPTION Dark brown, paler below, with greyish hindwing streaks and spots (less conspicuous above). Male with a pale streak (brand) above forewing dorsum. The rarer **Striped Black Crow** *Euploea doubledayi* (Thailand and northern Peninsular Malaysia) is larger, and the similar **Long-branded Blue Crow** *Euploea algea* (Thailand and Peninsular Malaysia) has a blue-glazed forewing upperside and longer brand. DISTRIBUTION S Burma to S Vietnam, southern and eastern areas of Thailand, Peninsular Malaysia, Singapore, Sumatra, Java, Borneo, Palawan and the Lesser Sunda Islands.
SUBSPECIES Represented by a single, variable subspecies, *gardineri*, in the region.
HABITS AND HABITAT Disturbed and primary inland forest at low to moderate elevations. Flies slowly at moderate height.

Striped Blue Crow
■ *Euploea mulciber* 5.0cm

DESCRIPTION Male upperside black-brown with a blue-glazed forewing, spotted bluish white. Underside lighter brown with short, buff streaks and spots. Female paler, with more striking bluish-white forewing spots, and with long buff hindwing streaks. The male **Blue-branded King Crow** *Euploea eunice* is larger with rounder wings and a light blue streak above the upperside forewing dorsum. The **Dwarf Crow** *Euploea tulliolus* has a similar wing shape but is small. Both occur in Thailand and Peninsular Malaysia.
DISTRIBUTION India to S China, through Thailand, Peninsular Malaysia, Singapore, Sumatra, Java, Borneo and the Philippines.
SUBSPECIES A single subspecies, *mulciber*, occurs in the region.
HABITS AND HABITAT Common in disturbed and primary forest at all elevations. Flies slowly at low to moderate height, often visiting flowers at the edges of forest roads.

TOP: *Male*. ABOVE: *Female*

Great Crow ▪ *Euploea phaenareta* 6.0cm

DESCRIPTION The largest crow. Brown with purplish-white streaks and spots on the outer areas of both wings. The **Malayan Crow** *Euploea camelzeman* is a large and similar-looking crow that occurs up to moderate elevations in Thailand and Peninsular Malaysia. It is darker with whiter and better defined wing spots.

Female of ssp. castelnaui

DISTRIBUTION Sri Lanka, Burma, Peninsular and SE Continental Thailand, S Indo-China, Peninsular Malaysia, Singapore, the Indonesian Islands and Palawan, eastwards to Australia and the Bismarck Islands. SUBSPECIES The race *castelnaui* occurs as far north as S Thailand, and the less strongly spotted race *drucei* in the southeast.
HABITS AND HABITAT Slow-flying. Commonest in coastal forest but occurs in inland forest in the lowlands. May sometimes occur in villages and towns where its host plants Sea Mango *Cerbera manghas* and Pong-Pong *C. odollam* are grown.

Magpie Crow
▪ *Euploea radamanthus* 4.0cm

Male

DESCRIPTION Male with wings blue-glossed black, with light blue peripheral spots, and a white patch at the forewing costa and hindwing dorsum. Female brown with more extensive white forewing bar and white hindwing streaks. The female is mimicked by the female of the Courtesan (p.93) and both sexes by forms of the Great Blue Mime (p.26).
DISTRIBUTION Sikkim eastwards to Indo-China, through Thailand, Peninsular Malaysia, Singapore, Sumatra, Java, Borneo and Palawan.
SUBSPECIES Occurs as a single ssp. *radamanthus* from Thailand to Singapore.
HABITS AND HABITAT Flies slowly. Common in inland forest in the lowlands and highlands. Males are commonly seen along roadsides close to forest and may settle on damp ground.

> **Browns – Elymniini, Zetherini, Melanitini (Satyrinae)**
> Butterflies of the subfamily Satyrinae that utilise grasses, bamboos, palms and orchids as host plants. Adults often inhabit bushland or the forest understorey and feed on rotting fruit.

Common Evening Brown
■ *Melanitis leda* 4.0cm

DESCRIPTION Upperside brown with white-centred, black spots on the forewing apex and hindwing tornus. Underside with fine dark lines forming a variable pattern, and with marginal eyespots (very small in the dry season). The rarer and larger **Great Evening Brown** *Melanitis zitenius* is reddish brown above. The **Dark Evening Brown** *Melanitis phedima* is darker, with spots almost absent. Both occur in Thailand and Peninsular Malaysia.

Wet season form

DISTRIBUTION Tropical Africa to Japan, through Southeast Asia, Australia and the Bismarck Archipelago.
SUBSPECIES Represented by ssp. *leda*.
HABITS AND HABITAT Common in most lowland habitats. Low-flying. Most active in the evening. Its host plants are species of grasses, bamboos and cereals such as rice.

Dry season form

> **Palmflies – Elymnias**
> Fifteen species in the region. Many are very rare. All mimic milkweed butterflies (p.47) or jezebels (pp.35–6).

Tawny Palmfly ■ *Elymnias panthera* 4.0cm

DESCRIPTION Brown with a black-spotted, buff-coloured hindwing costal band, and a touch of buff on the forewing tornus. Underside with numerous small reddish-brown streaks. Pale areas and white-centred black spots on the underside correspond with upperside markings.
DISTRIBUTION S Thailand, Peninsular Malaysia, Singapore, Sumatra, Java, Borneo and Palawan.
SUBSPECIES Two regional subspecies: *panthera* (mainland) and *tiomanica* (Tioman Island).
HABITS AND HABITAT Usually flies in shady areas near palm trees in disturbed and primary forest in the lowlands.

Ssp. panthera

Common Palmfly ■ *Elymnias hypermnestra* 4.0cm

DESCRIPTION Underside mottled reddish brown, usually with a white costal hindwing spot and sometimes pale patches. Males blue-black above, forewing streaked pale blue. Females of the southern sspp. *agina* and *nimota* are paler with larger and lighter blue

markings. Females of *meridionalis*, *tinctoria* and *discrepans* have orange and white areas, resembling the Plain Tiger (p.47).

DISTRIBUTION India to SE China, Thailand, Peninsular Malaysia, Singapore, Sumatra, Java, the Lesser Sunda Islands, Borneo and the Philippines.

SUBSPECIES Eastern Continental Thailand: *meridionalis*; N, W and S Thailand, and NW Peninsular Malaysia including Langkawi: *tinctoria*; South Kedah and Penang Island: *discrepans*; the rest of Peninsular Malaysia south to Singapore: *agina*; islands of Tioman and Aur: *nimota*.

HABITS AND HABITAT Common in gardens, villages, oil-palm plantations and the forest edge in the lowlands. Flies fairly slowly, landing often. Breeds on a variety of palms, and adults often fly near their host plants.

Male of ssp. agina

Ssp. penanga

Pointed Palmfly

■ *Elymnias penanga* 3.5cm

DESCRIPTION Differs from the Common Palmfly in having narrower and more pointed forewings. Male resembles the Common Palmfly above, but the female occurs in different forms. It may be similar to the male but paler, or greenish brown with a white forewing band or white hindwing patch.

DISTRIBUTION NE India to Indo-China, Peninsular Malaysia, Singapore (very rare), Sumatra and Borneo.

SUBSPECIES Occurs as ssp. *penanga* from Singapore to S Thailand, and as *chelensis* to the north.

HABITS AND HABITAT Similar to the Common Palmfly, but less common, favouring denser vegetation, including primary forest.

TREE-BROWNS AND FORESTERS – *LETHE*

Brown butterflies with cream or orange-brown markings and underside eyespots. Hindwing usually with a short, triangular tail. Twenty species occur in the region, of which 12 do not occur south of Thailand. Most are rare. They fly near the ground and rest among dead leaves. More active at dawn and dusk.

Bamboo Tree-Brown ■ *Lethe europa* 3.3cm

DESCRIPTION Underside dark brown with a white forewing band, a narrow white line across both wings, and irregular marginal eyespots. Upperside brown with the forewing tip spotted, and a white forewing bar in the female. The somewhat similar **Banded Tree-Brown** *Lethe confusa* (Thailand and Peninsular Malaysia) has rounded marginal eyespots. DISTRIBUTION India to China, Thailand, Peninsular Malaysia, Sumatra, Java to Sulawesi, Borneo and the Philippines. The only species of *Lethe* occurring in Singapore. SUBSPECIES Occurs as ssp. *malaya* from Singapore to S Thailand and as *niladana* in the rest of Thailand. HABITS AND HABITAT Uncommon, occurring mainly in villages and bamboo glades.

Female of ssp. malaya

Common Red Forester ■ *Lethe mekara* 3.4cm

DESCRIPTION Distinguishing characters: brown, with the most basal dark band straight and outwardly edged whitish, and with some of the marginal spots distorted rather than round. Female with white forewing markings. The **Angled Red Forester** *Lethe chandica* (Thailand and Peninsular Malaysia) is darker with a wavy dark band. DISTRIBUTION NE India to S China, Thailand, Peninsular Malaysia, Sumatra, Borneo and Palawan. SUBSPECIES Two

Female of ssp. gopaka

subspecies: *gopaka* (Peninsular Malaysia to S Thailand) and *crijnana* (the rest of Thailand). HABITS AND HABITAT Locally common in forest with bamboos.

Male of ssp. gopaka

Black-spotted Labyrinth
■ *Neope muirheadi* 3.5cm

DESCRIPTION Underside variegated shades of brown with a pale buff line across both wings and a series of black-centred marginal eyespots. Upperside brown with black-centred pale rings along the wing margins, which are smaller and fewer on the forewing of the male.

DISTRIBUTION N Burma, N Thailand and northern Indo-China, to China and Taiwan.

SUBSPECIES Occurs as ssp. *bhima* in N Thailand.

HABITS AND HABITAT Fast but low flying, landing frequently. Uncommon in forests up to moderate elevations. Keeps to shady areas near bamboos.

Tiger Brown
■ *Orinoma damaris* 3.8cm

DESCRIPTION Black with broad, creamy white markings between the wing veins. The creamy markings comprise streaks that break into detached spots in the outer areas of the wings. The forewing has a characteristic triangular orange spot at the base of the cell on both sides of the wing. This orange spot bears 2 prominent, black spots.

DISTRIBUTION N India to N Thailand, N Indo-China and S China.

SUBSPECIES The ssp. *damaris* is found in N Thailand.

HABITS AND HABITAT Flies slowly and close to the ground in forest clearings or in the understorey, mimicking the flight of species of milkweed butterflies (p.47) that it resembles. Restricted to montane forests.

Ssp. damaris, Northeast India

Malayan Owl ▪ *Neorina lowii* 5.5cm

DESCRIPTION Large, with tails. The upperside and paler underside are dark brown with a cream-coloured hindwing apical patch and a smaller adjacent patch on the forewing costa. Upperside with 2 eyespots; underside with 3. Forewing with a series of small, white spots. The **Tailed Yellow Owl** *Neorina crishna*, with a yellow forewing-bar, occurs in S Thailand, and the **White Owl** *Neorina patria*, with a shorter tail and whiter wing bar, occurs in N Thailand.

DISTRIBUTION S Thailand, Peninsular Malaysia, Sumatra, Borneo and Palawan.
SUBSPECIES Represented by ssp. *neophyta*.
HABITS AND HABITAT Fast, undulating flight. Often rests in the cover of vegetation, but males come to wet ground. Inhabits forested areas up to moderate elevations, generally occurring locally near streams with bamboo thickets.

Blue Kaiser

▪ *Penthema darlisa* 6.5cm

DESCRIPTION Medium-large, occurring in forms that are considered subspecies but are of uncertain status. Upperside dark brown or black, sometimes with a blue gloss on the forewing. Hindwing termen with cream coloured arrowhead- and spot-shaped markings (ssp. *mimetica*), or with additional streaks and spots on both wings (ssp. *melema*) or on only the forewing, but having broad yellow stripes on the hindwing (ssp. *merguia*). All forms more chestnut brown beneath, marked almost as in upperside.

DISTRIBUTION N India to Thailand, Indo-China and S China.
SUBSPECIES N Thailand: *melema*; S Thailand: *merguia*; the rest of Thailand: *mimetica*.
HABITS AND HABITAT Slow-flying, at low to moderate height. Inhabits forests and the forest edge. Males sometimes come to puddles.

ABOVE RIGHT AND RIGHT:
Male of ssp. melema

BUSH-BROWNS – MYCALESIS
Twenty-nine species in the region. Small, brown or orange-brown butterflies. Underside with eyespots along the outer wing margins, usually with pale or dark lines across both wings, and sometimes with a pale bar across the forewing apex. They fly moderately fast but low, often close to the ground, resting frequently but briefly.

Male

Tawny Bush-Brown
▪ *Mycalesis anapita* 2.0cm

DESCRIPTION Underside orange-brown with 2 darker rusty-brown lines spanning across both wings and eyespots along the outer wing margins. Upperside deep orange-brown with a black forewing border from apex to tornus. **Eliot's Bush-Brown** *Mycalesis patiana*, found in Peninsular Malaysia, is very similar but darker, and the black upperside forewing eyespot touches the black border. It occurs in Peninsular Malaysia. Females of both species are larger and paler.
DISTRIBUTION S Burma, S Thailand, Peninsular Malaysia, Sumatra and Borneo. Absent from Singapore.
SUBSPECIES Occurs as ssp. *anapita* in this region.
HABITS AND HABITAT Uncommon and local in occurrence, inhabiting inland forest up to submontane elevations. Faster flying than most bush-browns.

Male

Malayan Bush-Brown
▪ *Mycalesis fusca* 2.1cm

DESCRIPTION Differentiated from the somewhat similar Tawny Bush-Brown and Eliot's Bush-Brown by its duller orange-brown colour, less strongly white-centred eyespots, and uniform, dark brown upperside. The female is larger and paler.
DISTRIBUTION S Burma, S Thailand, Peninsular Malaysia, Singapore, Nias, Sumatra and Borneo.
SUBSPECIES The subspecies in this region is *fusca*.
HABITS AND HABITAT Common, especially in grassy and bushy verges of logging tracks and roads in lowland forests.

Mottled Bush-Brown
■ *Mycalesis janardana* 2.3cm

DESCRIPTION Dark brown; underside with a buff white line across both wings. Distinguished by a full series of uniformly sized eyespots, and a densely dark-streaked underside. The browner **Purple Bush-Brown** *Mycalesis orseis* (S Thailand to Singapore) has less evenly sized spots, a wider band with a few dark streaks. The male is purple-washed above.
DISTRIBUTION S Thailand, Peninsular Malaysia, Sumatra, Borneo, Palawan and Java eastwards to Sulawesi and the Moluccas.
SUBSPECIES Occurs as a single subspecies, *sagittigera*, in this region.
HABITS AND HABITAT Inhabits grassy areas at the edges of forests and in forest clearings in the lowlands, and can be locally abundant.

Male

Dark-branded Bush-Brown ■ *Mycalesis mineus* 2.4cm

DESCRIPTION One of a group of similar-looking bush-browns characterised by unequal-sized marginal eyespots and a straight, white line across both wings. Black-brown, with the white line that outlines the inner edges of the hindwing eyespots sharply inflected

above its midpoint. In the paler **Long-branded Bush-Brown** *Mycalesis visala*, the line is more evenly inflected. The **Dingy Bush-Brown** *Mycalesis perseus* differs from both by having the 4 lowermost, hindwing eyespots out of line. Both occur from Thailand to Singapore.
DISTRIBUTION India and Sri Lanka to S China and throughout Southeast Asia.
SUBSPECIES The ssp. *macromalayana* occurs from Singapore to S Thailand and ssp. *mineus* in the rest of Thailand.
HABITS AND HABITAT Common, inhabiting disturbed and primary forest, often at forest edges. May also occur in villages, parks and large gardens.

Male of ssp. macromalayana

Male of ssp. cinerea

Dark Grass-Brown

■ *Orsotriaena medus* 2.2cm

DESCRIPTION Resembles a bush-brown with its dark brown colour, plain upperside, and its underside pattern. Distinguished on the underside by its whiter wing stripes and fewer, relatively large eyespots.

DISTRIBUTION India and Sri Lanka to S China, through Southeast Asia to New Guinea and Australia.

SUBSPECIES Occurs as ssp. *medus* in most of Thailand and as the very similar ssp. *cinerea* in southernmost Thailand, Peninsular Malaysia and Singapore (sometimes considered to be the same subspecies as *medus*).

HABITS AND HABITAT Common in a wide range of habitats in the lowlands wherever there are grassy areas, including forest edges, villages, parks and sometimes gardens.

Angled Cyclops ■ *Erites angularis* 2.8cm

DESCRIPTION One of 5 species of cyclopes in the region. All are medium brown above with faint to clear eyespots. Beneath they are grey, with fine dark streaks, a characteristic

double orange-brown line and orange-ringed eyespots. The forewing tornal eyespot may be large and prominent. The Angled Cyclops is recognised by its sharply bent inner orange line on the hindwing. The **Eyed Cyclops** *Erites argentina*, which occurs in S Thailand and Peninsular Malaysia, has a small eyespot between the largest forewing spot and the triple eyespot nearer the apex.

DISTRIBUTION Burma, Thailand, Laos, Peninsular Malaysia and Sumatra. Absent from Singapore.

SUBSPECIES The subspecies in the region is *angularis*.

HABITS AND HABITAT Local in occurrence and generally rare, occurring only in lowland dipterocarp forest. Feeble in flight, keeping low in the forest understorey.

Blue Catseye ■ *Coelites epiminthia* 3.3cm

DESCRIPTION Upperside deep purple-blue with black-brown outer borders. Underside dark brown, with dark and pale discal lines and subdued marginal eyespots. Males have a black tornal patch on the hindwing upperside. The **Purple-streaked Catseye** *Coelites euptychioides* (S Thailand and Peninsular Malaysia) has the purple-blue colouration confined to the hindwing costa and tornus. The **Scarce Catseye** *Coelites nothis* (Continental Thailand) has the hindwing eyespots clearly ringed with orange.

DISTRIBUTION S Burma, S Thailand, Peninsular Malaysia, Sumatra, Java, Borneo and Palawan.

SUBSPECIES The ssp. *epiminthia* occurs in this region.

HABITS AND HABITAT Uncommon and usually noticed as a flash of blue when it takes to flight. Inhabits inland forest in the lowlands. Keeps to dense forest, flying low and resting in undergrowth.

Male

Brown-banded Ringlet
■ *Ragadia makuta* 2.3cm

DESCRIPTION Pale, unmarked brown above, and creamy buff below with dark brown stripes on both wings and a series of silver-centred eyespots along the wing margins. Two species that differ in having creamy white bands on the upperside are the **Zebra Ringlet** *Ragadia critolaus* (W Thailand and central highlands of Peninsular Malaysia) and **Striped Ringlet** *Ragadia crisilda* (Thailand and NW Peninsular Malaysia). The latter has the outer white band completely dark-dusted.

DISTRIBUTION S Thailand, Peninsular Malaysia, Sumatra, Java, Borneo and Palawan. Absent from Singapore.

SUBSPECIES The local subspecies is *siponta*.

HABITS AND HABITAT Occurs in inland forest, especially in the lowlands. Slow and feeble in flight, keeping near the ground in shaded areas of heavy forest.

Male

RINGS – YPTHIMA
Small grey-brown butterflies, mottled beneath with fine dark streaks, and with yellow-ringed eyespots. They favour grassy areas and bushland, flying low and sometimes basking in the sun with wings open. Fifteen species occur in the region. All are found in Thailand and a proportion occur in Peninsular Malaysia and Singapore. The species status of some forms is still uncertain. Females are larger with rounder wings.

Common Four-Ring
■ *Ypthima huebneri* 1.7cm

DESCRIPTION Characterised by its small size and 4 eyespots or rings on the hindwing underside (the joined double eyespot at the tornus counted as 1).
DISTRIBUTION Sri Lanka and India to S China, through Thailand, Peninsular Malaysia and Singapore.
SUBSPECIES No subspecies have been named; it is similar throughout its range.
HABITS AND HABITAT Very common in grassy areas in vacant land, parks, fields, villages, roadsides and even urban gardens in the lowlands.

Common Five-Ring ■ *Ypthima baldus* 1.8cm

DESCRIPTION The commonest of a group of confusingly similar rings that have 5 eyespots (the double tornal eyespot counted as 1). The very similar **Horsfield's Five-Ring** *Ypthima horsfieldii* (S Thailand and Peninsular Malaysia) inhabits wooded areas and has smaller, more widely separated and more evenly sized eyespots.
DISTRIBUTION Sri Lanka and India to China, Taiwan, Korea, Japan and E Russia, through Thailand, Malaysia, Singapore, and the Indonesian Islands as far as Sulawesi.

Ssp. newboldi

Ssp. newboldi

SUBSPECIES Occurs as ssp. *newboldi* from Singapore to S Thailand and as ssp. *baldus* in the rest of Thailand.
HABITS AND HABITAT Common in parks, villages and wayside vegetation in the lowlands.

Common Three-Ring

■ *Ypthima pandocus* 2.3cm

DESCRIPTION Easily recognised by being larger than most common rings and by the presence of only 3 eyespots on the hindwing underside.
DISTRIBUTION S Thailand, Peninsular Malaysia, Singapore, Sumatra, Java, Borneo, Palawan and the Philippines.
SUBSPECIES Occurs as ssp. *corticaria* from S Thailand to Singapore, but a darker ssp. *tahanensis* with smaller spots occurs around the summit of Gunung Tahan.
HABITS AND HABITAT Common on the mainland at all elevations in coastal and inland areas. Usually seen at the edges of forests, along forest roads and tracks, and in waysides, villages, parks and gardens.

Ssp. corticaria

Ssp. corticaria

Yellow-barred Pan ■ *Xanthotaenia busiris* 3.5cm

DESCRIPTION Upperside rich, rusty brown turning black towards the apical half of the forewing, and with a prominent yellow bar across the forewing running from mid-costa to near tornus. Underside paler rusty brown with the yellow bar visible and with several grey-centred discal eyespots and a few wavy, dark rusty-brown lines. Females are larger and paler. Can be confused with the Tufted Jungle King (p.71), especially in flight, but the latter is more strongly marked and is much larger.
DISTRIBUTION S Burma, S Thailand, Peninsular Malaysia, Sumatra and Borneo. Extinct in Singapore.
SUBSPECIES The ssp. *busiris* occurs in this region.
HABITS AND HABITAT Fairly common in inland forest, mainly in the lowlands. It is usually seen flying close to the forest floor along forest paths; strongly attracted to fallen forest fruits, on which it feeds.

Common Faun ■ *Faunis canens* 3.5cm

DESCRIPTION Upperside uniformly yellowish red-brown. Underside brown with dark lines and small white spots running down both wings. Sexes alike. In the **Dusky Faun**

Faunis kirata (S Thailand and Peninsular Malaysia), the dark lines are wider, and the hindwing darkened beneath in the male, obscuring the dark lines.

DISTRIBUTION NE India to S China, Thailand, Peninsular Malaysia, Singapore, Sumatra, Java, Borneo and Palawan.

SUBSPECIES A single ssp. *arcesilas* occurs in the region.

HABITS AND HABITAT Inhabits inland forest up to submontane elevations. Keeps to forest cover and flies moderately fast near the ground, often resting on the forest floor. Fond of fallen forest fruits.

Graceful Faun ■ *Faunis gracilis* 3.0cm

DESCRIPTION Smaller and yellower than other fauns, especially beneath. The dark underside lines are more contrasting, and 2 of the white spots on the hindwing are ringed, forming small eyespots. Females are larger. The **Large Faun** *Faunis eumeus* (N and NE Thailand) is large with prominent yellow spots in place of white underside spots. The female is diffusely yellow-banded on the forewing upperside.

DISTRIBUTION S Thailand, Sumatra, Peninsular Malaysia and Borneo.

SUBSPECIES None. Uniform in appearance across its range.

HABITS AND HABITAT Similar to the Common Faun but may be rare in some locations and common in others.

Pallid Faun

■ *Melanocyma faunula* 4.5cm

DESCRIPTION The only species in its genus. Moderately large. Pale grey-brown with the hindwing dorsum and tornus yellow, and with prominent black, zigzagging lines on the underside that make the butterfly unmistakable. Males have a small black patch near the tornus.

DISTRIBUTION Burma, S and SE Thailand, and Peninsular Malaysia.

SUBSPECIES Occurs as 2 subspecies, *faunula* in Peninsular Malaysia and S Thailand, and *kimurai* in SE Thailand.

HABITS AND HABITAT Local in occurrence, inhabiting inland forest. Commoner in the highlands. Flies rather slowly in the canopies of small understorey trees and is usually seen at the forest edge, but does not venture far from thick forest.

Ssp. faunula

Burmese Junglequeen ■ *Stichophthalma louisa* 7.0cm

DESCRIPTION The least rare of 4 very large species of junglequeens in Thailand. The Burmese Junglequeen is ochreous brown above with a white forewing band and, like virtually all junglequeens, has black arrow-like markings along the outer wing margins. On the underside it is paler orange-brown with black transverse lines and black-ringed, white-centred, red-brown discal eyespots.

DISTRIBUTION NE India, Burma, Thailand and Vietnam.

SUBSPECIES Four subspecies from this region: *louisa* in NW and W Thailand, *mathilda* in N Thailand, *siamensis* in SE Thailand and *ranohngensis* in S Thailand.

HABITS AND HABITAT Flies slowly in the understorey, with a gliding zigzag flight path. Lands on vegetation with wings closed. Rare in forests up to the highlands. Keeps to forest cover, and is more commonly seen along forest gullies and streams.

Ssp. louisa

Palm King
▪ *Amathusia phidippus* 5.0cm

DESCRIPTION Pale brown below, with the typical dark brown band, whitish transverse streaks and 2 widely separated hindwing eyespots. Brown above with orange-brown forewing apical borders. Females are paler above with yellower borders.
DISTRIBUTION Parts of India and throughout Southeast Asia.
SUBSPECIES The ssp. *phidippus* occurs in this region.
HABITS AND HABITAT Although very difficult to distinguish from other species of palm kings in the field, it is the most likely species to be seen, being common in coconut and oil-palm plantations and also in villages and forests in the lowlands wherever these palms that are its host plants are grown nearby.

Female of ssp. dilucida

Kohinoor ▪ *Amathuxidia amythaon* 5.5cm

DESCRIPTION Male black-brown above with a wide, blue band across the forewing; female paler brown with an orange-brown band. Underside greyish brown with dark lines across the wings and two faint eyespots. Hindwing with a lobe-like tail.
DISTRIBUTION N India and throughout Southeast Asia.
SUBSPECIES Occurs as ssp. *dilucida* in Peninsular Malaysia and S Thailand, *annamensis* in SE Thailand and *amythaon* in the rest of Thailand.
HABITS AND HABITAT Prefers dense primary dipterocarp forest in the lowlands. Uncommon and secretive, keeping still among thick vegetation and making short flights into denser forest when disturbed.

Saturn ■ *Zeuxidia amethystus* 5.0cm

DESCRIPTION Large, with a short hindwing tail. Male black-brown above with a blue forewing band and hindwing tornus. Female tawny brown with a broken, pale yellow forewing band. Underside shades of brown, yellower and more contrastingly marked in the female, leaf-like, with a dark central line and with 2 hindwing eyespots. In the **Scarce Saturn** *Zeuxidia doubledayi* (Peninsular Malaysia to S and SE Thailand) the blue on the hindwing of the male extends up the termen, and the pale forewing band of the female is purplish white.

DISTRIBUTION S Thailand, Peninsular Malaysia, Singapore, Sumatra, Borneo and the Philippines.

SUBSPECIES The regional subspecies is *amethystus*.

HABITS AND HABITAT Moderately fast and erratic flight. Keeps to the understorey in dense forest in the lowlands and highlands. Not rare, but rarely seen because of its cryptic habits.

Male

Great Saturn
■ *Zeuxidia aurelius* 7.0cm

DESCRIPTION One of the largest butterflies in the region. Differentiated from other saturns by its large size and whitish underside. Male with a very broad, blue band across the forewing upperside and some blue at the hindwing tornus. Female brown above with prominent white streaks and spots across the forewing, and with some on the hindwing.

DISTRIBUTION S Burma, S Thailand, Peninsular Malaysia, Sumatra and Borneo.

SUBSPECIES Represented locally by ssp. *aurelius*.

HABITS AND HABITAT Similar to the Saturn but rarer. Very conspicuous when it takes to flight. Usually seen only when disturbed along deep forest paths.

Female

Blue-banded Jungle Glory ▪
Thaumantis odana 5.0cm

DESCRIPTION Medium-large, without pointed wings or a tail. Upperside brown with a bright blue band across the forewing. Underside with contrasting brown and buff bars and 2 hindwing eyespots. The **Jungle Glory** *Thaumantis diores* occurs in Thailand and is easily recognised by its rounder wings and bright blue, oval forewing band and hindwing patch.

DISTRIBUTION S Thailand, Peninsular Malaysia, Sumatra, Borneo and Java.

SUBSPECIES Occurs as ssp. *pishuna* in this region.

HABITS AND HABITAT Inland forest in the lowland and highlands. Flies close to the forest floor, resting among dead leaves on the ground or on low plants.

Dark Blue Jungle Glory
▪ *Thaumantis klugius* 4.5cm

DESCRIPTION Male dark, gleaming blue above with black-brown borders. Female brown above with purple wing bases and pale spots and lines on the forewing. Underside variable, similar to the Blue-banded Jungle Glory but less contrastingly marked. The **Dark Jungle Glory** *Thaumantis noureddin* has a more quadrate forewing tip and more produced hindwing. Above the male is black-brown, with inconspicuous blue at the wing bases, and the female is marked as in the female Dark Blue Jungle Glory. The Dark Jungle Glory occurs from S Thailand to Peninsular Malaysia.

DISTRIBUTION S Burma, S Thailand, Peninsular Malaysia, Singapore, Sumatra and Borneo.

SUBSPECIES The local subspecies is *lucipor*.

HABITS AND HABITAT Similar to the Blue-banded Jungle Glory but inhabits denser forest in the lowlands, and is less tolerant of forest disturbance.

Tufted Jungle King ■ *Thauria aliris* 6.0cm

DESCRIPTION Large, without tails. Upperside black, with a cream-coloured forewing bar. Forewing base and hindwing tornus and apex orange-brown. Underside grey, chestnut brown and white, with 2 prominent hindwing eyespots. Male has a hindwing hair tuft. Can be confused with the Yellow-barred Pan, which is much smaller. The smaller **Jungle King** *Thauria lathyi*, in which the male lacks a hindwing tuft, occurs in N and SE Thailand.
DISTRIBUTION NE India, S Burma, much of Thailand, Peninsular Malaysia and Borneo.
SUBSPECIES N Thailand: ssp. *intermedia*; S Thailand and Peninsular Malaysia: *pseudaliris*.
HABITS AND HABITAT Inhabits the understorey of inland forest in the lowlands. Has a swifter flight than most owls. Rests on the ground or amongst understorey vegetation.

Ssp. pseudaliris

Common Duffer
■ *Discophora sondaica* 4.0cm

DESCRIPTION Medium-large, with pointed wings; no tail. Mottled brown beneath with a dark central line and 2 hindwing eyespots. Male black-brown above with bluish white forewing spots. Female paler with 3 rows of pale brown spots. The larger **Great Duffer** *Discophora timora* (Thailand and Peninsular Malaysia) has small, white forewing spots in the male and a yellowish forewing band in the female.
DISTRIBUTION N India to S China and Southeast Asia up to the Philippines and Bali.
SUBSPECIES Two regional subspecies: *despoliata* (Singapore to S Thailand), and *zal* (to the north).
HABITS AND HABITAT Usually seen near bamboo thickets at low to moderate elevations. Flies fast at moderate height, resting close to dead leaves on understorey plants.

Male of ssp. despoliata

BEAKS – LIBYTHEINAE
A small subfamily of butterflies in the family Nymphalidae. They have protruding beak-like palps and curved, rectangular forewing tips.

Club Beak ▪ *Libythea myrrha* 2.5cm

DESCRIPTION Ground colour black-brown above. Forewing with an expanding orange streak from forewing base to beyond the middle of the wing, and a few subapical orange spots. Hindwing with an orange band from dorsum to near the apex. Underside grey and

brown, and heavily mottled. The **Common Beak** *Libythea celtis* (N Thailand) has the forewing orange streak disjointed and the subapical spots mostly whitish.
DISTRIBUTION Sri Lanka and N India to W China, through Thailand, Peninsular Malaysia, Sumatra, Java, Borneo and Palawan.
SUBSPECIES Occurs as ssp. *hecura* up to S Thailand, and as ssp. *sanguinalis* further north.
HABITS AND HABITAT Very localised inhabitant of inland forest at moderate elevations, often near rivers. Flies low but fast, landing with closed wings. Males settle on wet ground.

Male of ssp. hecura

White-spotted Beak ▪ *Libythea narina* 2.5cm

DESCRIPTION Wings brown above, with yellowish-white spots, including one on the hindwing costa. Underside contrasting brown and white, and heavily mottled. The **Blue**

Beak *Libythea geoffroy* (Thailand) has a bluish-purple wash above, which is more extensive in the male.
DISTRIBUTION NE India to Vietnam, through Thailand, northern Philippines and most of the Southeast Asian islands as far as New Guinea. Present in Peninsular Malaysia only on Langkawi Island, and absent from Singapore and Borneo.
SUBSPECIES The local subspecies is *rohini*.
HABITS AND HABITAT Similar to the Club Beak.

Male

> ### Leafwings – Charaxinae
> A subfamily of stout-bodied Nymphalids. Nawabs (*Polyura*) and rajahs (*Charaxes*) have pointed forewing tips and short, pointed hindwing tails, while begums (*Prothoe* and *Agatasa*) have rounded forewings and hindwing tails. They feed on rotting fruit, and males also feed on carrion. Strong-flying, resting with wings closed.

Plain Nawab ▪ *Polyura hebe* 4.0cm

DESCRIPTION One of 5 smaller nawabs in the region. Yellowish white above with a broad, black outer forewing border and narrower hindwing border. Underside rich brown with a central greenish white band outlined by deeper brown. The **Common Nawab** *Polyura athamas* (Thailand and Peninsular Malaysia) has much narrower yellowish-white areas.
DISTRIBUTION S Burma, S Thailand, and Southeast Asia up to Palawan and the Lesser Sunda Islands.
SUBSPECIES Three local subspecies: *chersonesus* (S Thailand and Peninsular Malaysia), *takizawai* (Tioman Islands) and *plautus* (Singapore).
HABITS AND HABITAT Inhabits lowland forests and adjacent villages and parks.

Male of ssp. plautus

Jewelled Nawab ▪ *Polyura delphis* 5.0cm

DESCRIPTION Yellowish white above with the outer third of the forewing black. Underside silvery white with small black, blue-grey and pale orange markings, especially on the hindwing. The **Great Nawab** *Polyura eudamippus* (Thailand and Peninsular Malaysia) has wider, white-spotted black borders above and red-brown lines below, one of which is forked.
DISTRIBUTION NE India to Thailand, Peninsular Malaysia, Sumatra, Java, Borneo and Palawan.
SUBSPECIES Two local subspecies: *concha* (Peninsular Malaysia and S Thailand) and *delphis* (in the north).
HABITS AND HABITAT Uncommon in inland forest up to moderate elevations. Usually seen near rocky rivers.

Male of ssp. concha

Blue Nawab ■ *Polyura schreiber* 4.5cm

DESCRIPTION A large and distinctive nawab. Black above with blue-dusted wing bases and a white, central wing band that has light blue outer edging. Underside light brown with the white band inwardly bordered by a brown to greenish-brown stripe and outwardly bordered by crescent-shaped spots. Hindwing underside with orange marginal markings.

DISTRIBUTION India to Thailand, Peninsular Malaysia, Singapore, Sumatra, Java, Borneo, Palawan and the Philippines.

SUBSPECIES Ssp. *tisamenus* is found from Singapore to S Thailand, and ssp. *assamensis* in the rest of Thailand.

HABITS AND HABITAT Found in inland forest at low to moderate elevations, as well as in villages near forest. Its host plant is the Rambutan tree (*Nephelium lappaceum*).

Ssp. tisamenus

Tawny Rajah ■ *Charaxes bernardus* 4.3cm

DESCRIPTION One of 9 species of rajahs in the region, most of which are very rare. Male rich orange-brown above with the apical half of the forewing typically black, and sometimes with a white mid-forewing band in ssp. *hierax*. Underside brown with dark, wavy lines forming a scaly pattern. Female paler with a yellow-white mid-forewing band and more contrasting underside markings. The **Variegated Rajah** *Charaxes kahruba* (Thailand, mainly in the north) is highly variegated below and has narrower black areas above.

DISTRIBUTION Sri Lanka and India to China, Southeast Asia, New Guinea and the Bismarck Islands.

SUBSPECIES Represented by ssp. *crepax* in Peninsular Malaysia and S Thailand, and *hierax* north of this range.

HABITS AND HABITAT Uncommon, but may be seen along forest paths and roads in both the lowlands and highlands.

Male of ssp. crepax

Blue Begum ■ *Prothoe franck* 4.0cm

DESCRIPTION Black above with blue-dusting and a bright blue band across the forewing. Underside brown with contrasting dark markings and shining greenish-brown scaling in the outer third of the hindwing.

DISTRIBUTION NE India to Thailand, Peninsular Malaysia, Sumatra, Java, Borneo and the Philippines.

SUBSPECIES Three subspecies in the region: *uniformis* (Peninsular Malaysia and S Thailand), *angelica* (W Thailand) and *vilma* (other parts of Thailand). Northern subspecies have more white in the blue band.

HABITS AND HABITAT Inhabits forests, mainly at low to mid-elevations. Flies in the middle storey, and often seen as a flash of blue. Lands facing downwards on the trunks of small trees, where it is well camouflaged with wings closed.

Ssp. uniformis

Glorious Begum
■ *Agatasa calydonia* 5.0cm

DESCRIPTION A striking butterfly that is unmistakable. Light yellow on the forewing above with a broad, yellow-spotted black border. Light blue on the hindwing with a broad, black outer border. Underside buff and brown with prominent black, red and yellow streaks and spots, as well as touches of shining greenish brown and sometimes shining blue near the outer margins of the hindwing.

DISTRIBUTION Burma to Thailand, Peninsular Malaysia, Sumatra, Borneo and the Philippines.

SUBSPECIES The ssp. *calydonia* occurs from Peninsular Malaysia to S Thailand, and subspecies *belisama* further north.

HABITS AND HABITAT Similar to the Blue Begum but rarely seen unless it comes to the ground to feed on carrion or the droppings of animals.

Ssp. calydonia

LONGWINGS AND FRITILLARIES – HELICONIINAE
A subfamily of medium-sized Nymphalids. Some have elongated forewings. The hindwing margin may be toothed or wavy.

Male of ssp. hypsina

Malayan Lacewing
▪ *Cethosia hypsea* 4.3cm

DESCRIPTION Wing margins toothed. Orange-red above with black borders and a creamy forewing band. Female paler orange. Underside very ornate red, orange and white with many small black dots and stripes. In the paler **Northern Orange Lacewing** *Cethosia methypsea* (S Thailand to Singapore), the forewing bar is broken, as also in the even paler **Leopard Lacewing** *Cethosia cyane* (Thailand to Singapore). In the highland **Red Lacewing** *Cethosia biblis* (Thailand and Peninsular Malaysia), the band is absent and replaced by small spots.
DISTRIBUTION S Burma and S Thailand to Singapore and Java, and from Sumatra eastward to Palawan.
SUBSPECIES Two subspecies: *hypsia* (S Thailand to Singapore) and *elioti* (Islands of Tioman and Aur).
HABITS AND HABITAT Inhabits forests at all elevations. Fond of visiting flowers at the forest edge.

Male of ssp. hypsina

Tawny Coster ▪ *Acraea terpsicore* 3.0cm

DESCRIPTION Formerly known as *Acraea violae*. Deep orange above, with narrow, black outer borders and black wing spots. Female paler. The black thorax and black hindwing border are spotted white. Underside paler with markings more prominent.
DISTRIBUTION Sri Lanka and India to Vietnam, and has spread into Thailand and more recently into Peninsular Malaysia and Singapore.

Male

SUBSPECIES There are no subspecies. It is uniform in appearance throughout its range.
HABITS AND HABITAT Usually seen in gardens, parks, villages, roadsides, bushland and coastal scrub. Flies at moderate speed with a fluttering flight.

Male

Yellow Coster ■ *Acraea issoria* 3.0cm

DESCRIPTION Differentiated from the Tawny Coster by its darkened wing veins and paler, duller orange-yellow colour. Underside less spotted than in the Tawny Coster, and the dark scalloped lines on the hindwing margin are inwardly edged by an orange border. Female yellower and much more prominently dark-dusted, especially on the forewing.
DISTRIBUTION N India eastwards to S China, reaching N Thailand and S Burma. Also occurs in the hills of Sumatra, Java and Bali.

ABOVE AND BELOW: *Male*

SUBSPECIES Occurs as a single ssp. *sordice* in Thailand.

Female

HABITS AND HABITAT Inhabits open country and the forest edge in the mountains. Flight is slow and fluttering. Like the Tawny Coster, it is poisonous to predators.

Lesser Cruiser ■ *Vindula dejone* 4.3cm

DESCRIPTION Hindwing with a short, pointed tail. Male orange-brown above with dark wing edging and a narrow black line running down both wings. Hindwing with 2 small eyespots. Underside paler with lightly contrasting dark lines. Female differs in being grey with a white band running down both wings. The **Cruiser** *Vindula erota* (Peninsular Malaysia and Thailand) is larger and prefers higher altitudes in Peninsular Malaysia, although it occurs in the lowlands of Perlis, Langkawi Island and Thailand.
DISTRIBUTION S Thailand to Singapore, and throughout the Southeast Asian islands as far as the Moluccas.
SUBSPECIES Three subspecies: *erotella* (S Thailand to Singapore), *tiomana* (Tioman Island) and *rafflesi* (Aur Island).
HABITS AND HABITAT Inhabits disturbed and primary forest from lowlands to mid-elevations. Flies fast, but males come to wet ground where they may fly slowly, often landing with open wings.

Male of ssp. erotella

Female of ssp. erotella

Common Yeoman ■ *Cirrochroa tyche* 3.3cm

DESCRIPTION Eight species of yeoman (*Cirrochroa*) occur in the region. The Common Yeoman is orange-brown above with black lines and spots and a thin forewing border. On the underside it is pale orange-yellow with a uniform whitish stripe. Three other species with increasingly well-defined black forewing borders are the smaller **Little Yeoman** *Cirrochroa surya* (Thailand and NE Peninsular Malaysia), the **Plain Yeoman** *Cirrochroa*

malaya (Peninsular Malaysia) and the deeper orange **Malayan Yeoman** *Cirrochroa emalea* (S Thailand and Peninsular Malaysia).

DISTRIBUTION India to S China, Thailand, Peninsular Malaysia, Sumatra, Java, Borneo and Palawan.

SUBSPECIES Three regional subspecies: *rotundata* (S Thailand and Peninsular Malaysia), *mithila* (the rest of Thailand), and *aurica* (Tioman Islands), which has a white-marked brown female.

HABITS AND HABITAT Inhabits forests at all elevations. Flies fast, but males may settle on wet ground.

Male of ssp. rotundata

Banded Yeoman ■ *Cirrochroa orissa* 3.0cm

DESCRIPTION Differs clearly from the preceding species of yeoman in having a yellow band across the forewing. The forewing apex is black beyond the band. The equally

distinctive **Long-banded Yeoman** *Cirrochroa satellita* (S Thailand and Peninsular Malaysia) is deeper orange with black wing borders and a yellow-orange band running down both wings.

DISTRIBUTION S Burma, S Thailand, Peninsular Malaysia, Singapore, Sumatra and Borneo.

SUBSPECIES Occurs as ssp. *orissa* in this region.

HABITS AND HABITAT Similar to the Common Yeoman, but occurs more widely.

Male

Royal Assyrian
■ *Terinos terpander* 3.2cm

DESCRIPTION Male purple above with a dark, velvet brown patch on both wings, and purplish-white spots at the hindwing tornus. Underside densely striped with red-brown and grey wavy lines and bands. Female paler purple and brown above with a wavy purple line. Two distinctly larger species are the **Large Assyrian** *Terinos atlita* and **Assyrian** *Terinos clarissa*. The latter has the hindwing tornus orange-white. They occur in Thailand and Peninsular Malaysia.

Male of ssp. robertsia

DISTRIBUTION Thailand, Peninsular Malaysia, Singapore, Sumatra, Java, Borneo and Palawan.

SUBSPECIES Three regional subspecies: *robertsia* (Singapore to S Thailand), *intermedia* (the rest of Thailand) and *tiomanensis* (Tioman Island).

HABITS AND HABITAT Inhabits primary and secondary forest at all elevations. Flies short distances in the understorey, landing frequently on or under leaves.

Ssp. robertsia

Small Leopard ■ *Phalanta alcippe* 2.2cm

DESCRIPTION Rich orange-brown chequered with small black markings, and with a narrow black border. Underside paler orange-brown with darker lines and dark hindwing spots. The **Common Leopard** *Phalanta phalantha* (Thailand to Singapore), which is commoner in gardens and parks, is larger and paler with black wavy lines along the wing margins.

DISTRIBUTION Sri Lanka and India through Southeast Asia and the Papuan region as far east as the Solomon Islands. Absent from Singapore.

SUBSPECIES Four subspecies in the region: *alcippoides* (Thailand), *alcesta* (Peninsular Malaysia), *tiomana* (Tioman Island) and *aurica* (Aur Island).

HABITS AND HABITAT Commoner in the lowlands than the highlands, in forests and at the forest edge. Generally gentle in flight. Males come to wet sand.

Male of ssp. alcippoides

Vagrant ■ *Vagrans sinha* 3.0cm

DESCRIPTION Hindwing with a short pointed tail. Forewing narrow and somewhat pointed. Orange-brown above with a narrow black border, black forewing markings and small black hindwing spots. Underside mixed shades of purplish brown and dark and pale orange-brown, marked with disjointed dark lines, small black spots and whitish crescents.

DISTRIBUTION India to S China and throughout Southeast Asia. Absent from Singapore.

SUBSPECIES Ssp. *sinha* and *macromalayana* are found in Thailand and Peninsular Malaysia, respectively.

HABITS AND HABITAT Inhabits

Male of ssp. macromalayana

forested areas from the lowlands to highlands. Males are fond of feeding at damp spots in rocky areas. Fast-flying, and restless even when on the ground.

Ssp. macromalayana

Rustic ■ *Cupha erymanthis* 2.8cm

DESCRIPTION Wings rounded. Deep orange-brown above with a yellow band across the forewing, bordered by a black apex, and marked with black spots and black wavy lines. Underside paler yellow-brown with a pale yellow forewing band, and with irregular brown and white lines and black spots.

DISTRIBUTION Sri Lanka and India to S China, through Thailand, Sumatra, Peninsular Malaysia, Singapore, Java, the Lesser

Ssp. lotis

Sunda Islands, Borneo and Palawan.

SUBSPECIES The subspecies in the region are *lotis* (Thailand to Singapore) and *tiomana* (the islands of Tioman and Aur).

HABITS AND HABITAT Found in disturbed and primary inland forests at all elevations. Has a restless flight, settling frequently and briefly on bushes and small trees along forest paths and clearings.

Ssp. lotis

> **ADMIRALS AND RELATIVES – LIMENITIDINAE**
> A large subfamily of Nymphalids that includes big genera like sailers and barons. Usually fly with quick, intermittent wing flaps and wing glides, landing with wings open.

Clipper ▪ *Parthenos sylvia* 4.5cm

DESCRIPTION Black above with white forewing spots, and with blue and green stripes and lace-like markings. Creamy brown below with black marginal markings and chalky greenish-blue scaling on the hindwing.
DISTRIBUTION Sri Lanka and India to S China, through Southeast Asia and the Papuan region as far as the Bismarck Islands.
SUBSPECIES The ssp. *lilacinus* occurs in Peninsular Malaysia and S Thailand and the greener ssp. *apicalis* in the rest of Thailand.
HABITS AND HABITAT Occurs in inland forests in the lowlands and is usually seen in bushy clearings. Flies with quick wing-beats and fast glides.

Male of ssp. lilacinus

Knight ▪ *Lebadea martha* 2.9cm

DESCRIPTION Wings brown to orange-brown above with a series of white spots running down the forewing, forming a bar on the hindwing. Intricately marked with wavy, white lines and crescents and small black markings, especially in ssp. *parkeri*. Underside paler brown and less intricately marked. Female with whiter wing tips.
DISTRIBUTION India to S China, through Thailand, Peninsular Malaysia, Singapore, Sumatra, Java, Borneo and Palawan.
SUBSPECIES Three subspecies: *martha* (Thailand and the northwest of Peninsular Malaysia), *malayana* (the rest of Peninsular Malaysia) and *parkeri* (Singapore).
HABITS AND HABITAT Occurs in inland forest in the lowlands, often landing on low trees along forest paths.

Male of ssp. martha *(S Thailand)*

Green Commodore ◾ *Sumalia daraxa* 2.8cm

DESCRIPTION Dark brown above with a greenish-white wing-band running up both wings, breaking into spots towards the forewing apex. Hindwing with black marginal spots and a red tornal spot. Underside ground colour reddish brown. The **Malayan Commodore** *Sumalia agneya* (Peninsular Malaysia), which has similar habits, differs in having the forewing apical spots at an angle to the band.

DISTRIBUTION NE India to S China, through Thailand, Peninsular Malaysia, Sumatra and Borneo.

SUBSPECIES In Thailand, it is represented by ssp. *daraxa* and, in Peninsular Malaysia, by ssp. *belliatus*.

HABITS AND HABITAT A mountain species that is not easily seen due to its habit of flying fast and high in the canopies of small trees. However, males may descend onto rocky roads and paths.

Male of ssp. belliatus

Commander ◾ *Moduza procris* 3.3cm

DESCRIPTION Red-brown above with black markings and a band of pure white, largely adjoining spots running down both wings. The underside is more contrasting and the wing bases are greenish white.

DISTRIBUTION Sri Lanka and India to S China, through Thailand, Peninsular Malaysia, Singapore, Sumatra, Borneo, Palawan, Java and the Lesser Sunda Islands.

Male of ssp. milonia

SUBSPECIES Represented by three subspecies in this region:

procris (NW Peninsular Malaysia and most of Thailand), *milonia* (southeastern extreme of Thailand, the rest of Peninsular Malaysia, and Singapore) and *tioma* (Tioman Islands).

HABITS AND HABITAT Inhabits inland forest in the lowlands and is commoner in disturbed forest and at the edges of forest roads. Flies fast, but males may land on the ground.

Male of ssp. milonia

Colonel ■ *Pandita sinope* 2.8cm

DESCRIPTION Bright brownish orange above, with the basal third of the wings darkened by brown bands and the wing margins outlined by dark lines. An outwardly scalloped, narrow brown band also runs down the wings in the outer third. Underside paler.
DISTRIBUTION S and SE Thailand, Peninsular Malaysia, Singapore, Sumatra, Java, Borneo and Palawan.
SUBSPECIES Occurs as a single ssp. *sinope* in the region.
HABITS AND HABITAT Inhabits secondary vegetation and bushy areas near forest and rivers in the lowlands.

SERGEANTS – *ATHYMA*
Medium-sized butterflies with white or orange-brown spots on a dark ground colour. There are 15 species in the region. Some are confusingly similar. Differentiated from lascars (p.84) and sailers (p.85) by a stouter body and stronger flight.

Colour Sergeant ■ *Athyma nefte* 3.0cm

DESCRIPTION Unlike other sergeants, females differ in colour from the male. Male black above with bluish-white markings. Female brown above with either orange markings (female-form *neftina*) or brown markings (female-form *subrata*). Underside ground colour brown, paler in the female. The **Common Sergeant** *Athyma perius* (Thailand and Peninsular Malaysia) has rounder

Ssp. subrata, *female-form* neftina

wings and a brighter orange ground colour on the underside than other sergeants.

Male of ssp. subrata

DISTRIBUTION India to S China, through Thailand, Peninsular Malaysia, Singapore, Sumatra, Java, Borneo and Palawan.
SUBSPECIES The ssp. *subrata* occurs from Singapore to S Thailand, and ssp. *asita* occurs in the rest of Thailand.
HABITS AND HABITAT Found mainly in lowland forests, including logged forest. Fast-flying, but males may come to the ground.

Common Lascar ▪ *Pantoporia hordonia* 2.3cm

DESCRIPTION Upperside black with orange-brown streaks. Underside orange-yellow with darker markings and fine red-brown costal streaks. The very similar **Bornean Lascar** *Pantoporia sandaka* (Thailand and Peninsular Malaysia) is more orange and has a slightly wider orange marginal line on the forewing upperside. In the **Perak Lascar** *Pantoporia paraka* (Thailand to Singapore) and smaller **Little Lascar** *Pantoporia aurelia* (Thailand and Peninsular Malaysia) the fine, red-brown underside streaks are absent and, in the

latter species, the marginal band is wide.

DISTRIBUTION Sri Lanka and India to S China and Taiwan, Thailand, Peninsular Malaysia, Singapore, Sumatra, Borneo, Palawan, Java and the Lesser Sunda Islands.

SUBSPECIES A single ssp. *hordonia* occurs in the region.

HABITS AND HABITAT Inhabits inland forests in the lowlands. Has a gentle, flitting flight.

Male

Malayan Lascar ▪ *Lasippa tiga* 2.5cm

DESCRIPTION Species of *Lasippa* differ superficially from very similar *Pantoporia* in having better defined underside markings. The Malayan Lascar has the orange marginal line more strongly toothed at its midpoint than in the very similar **Burmese Lascar** *Lasippa heliodore* (W and S Thailand to Singapore).

The rare **Fuliginous Sailer** *Lasippa monata* (Thailand to Peninsular Malaysia) has brown instead of orange streaks.

DISTRIBUTION NE India to Indo-China, through Thailand, Peninsular Malaysia, Singapore, Sumatra, Java and Borneo.

SUBSPECIES Two subspecies are found in the region: *siaka* (Singapore to S Thailand), and *camboja* (E and SE Thailand and Langkawi Island).

HABITS AND HABITAT Similar to the Malayan Lascar but also inhabits bushland and overgrown plantations.

Male of ssp. siaka

SAILERS – *NEPTIS*
Similar in markings to sergeants but thinner-bodied and slower-flying. Usually black with white markings. Twenty-eight species occur in the region, many of which are difficult to differentiate in the field. They have a slow flitting flight and land with wings spread.

Common Sailer ■ *Neptis hylas* 2.7cm

DESCRIPTION Usually easily distinguished from other sailers by its more orange-brown underside with white markings outlined in black. In the similar-looking **Malayan Sailer** *Neptis duryodana* (S Thailand and Peninsular Malaysia), the underside is brown.

DISTRIBUTION Sri Lanka and India to S China, Thailand,

Peninsular Malaysia, Singapore, Sumatra, Borneo, Palawan, Java and the Lesser Sunda Islands. SUBSPECIES The two subspecies in the region are *papaja* (Singapore to S Thailand) and *kamarupa* (the rest of Thailand).

Ssp. papaja

HABITS AND HABITAT Common, especially in the lowlands, inhabiting a wide range of habitats including coastal and inland forest, wayside vegetation and villages.

Ssp. papaja

Burmese Sailer ■ *Neptis leucoporus* 2.9cm

DESCRIPTION Distinguished from other sailers by the white scaling around the base of the abdomen above. The underside ground colour is brown, and the lower mid-forewing white spots are narrower and more stepped than in the preceding species. There are a few species of sailers

with narrow, brown markings, among which the commonest is usually the **Chocolate Sailer** *Neptis harita* (Thailand to Singapore).

DISTRIBUTION S Burma, S and W Thailand, Peninsular Malaysia, Singapore, Sumatra, Java and Borneo.
SUBSPECIES The species is represented by ssp. *cresina* in this region.
HABITS AND HABITAT Common only in forested areas.

Male of ssp. goodrichi

Female of ssp. rayana

Banded Marquis
■ *Bassarona teuta* 4.0cm

DESCRIPTION Male black above with joined yellow-white spots forming a wing band. Underside pale brown. Female varies from being paler than the male (ssp. *teuta*) to being grey-brown with small white wedge-shaped spots on the forewing and sometimes hindwing (ssp. *goodrichi*). The **Redtail Marquis** *Bassarona recta* (Thailand and NW Peninsular Malaysia) has outwardly rounded white spots and a red spot at the tip of the hindwing. The female is paler.

DISTRIBUTION NE India to Indo-China, through Thailand, Peninsular Malaysia, Sumatra, Java, Borneo and Palawan.

SUBSPECIES Five subspecies: *teuta* (North and Central Thailand), *gupta* (SE Thailand), *goodrichi* (S Thailand and Peninsular Malaysia), *rayana* (Langkawi Islands) and *tiomanica* (Tioman Island).

HABITS AND HABITAT Inhabits heavy forest in the lowlands. Fast-flying, usually keeping to the canopies of low trees.

Great Marquis ■ *Bassarona dunya* 4.5cm

DESCRIPTION Relatively large. Brown above with small cream-coloured spots forming a broken line down both wings. Underside largely pale greenish blue. The **Redspot Duke**

Male

Dophla evelina (Peninsular Malaysia and Thailand) is somewhat similar looking, with similar habits, but lacks creamy spots, has a red spot in the forewing cell and has the forewing tip outcurved.

DISTRIBUTION S Burma, S Thailand (occasionally in the W and N), Laos, Peninsular Malaysia, Sumatra, Java, Borneo and Palawan.

SUBSPECIES There is a single ssp. *dunya* in this region.

HABITS AND HABITAT Confined to lowland forest. Fast-flying, usually keeping to the understorey, but fond of feeding on rotting fruit on the forest floor.

> ### BARONS – EUTHALIA
> Fast-flying, medium-sized butterflies with dissimilar sexes. Diverse, with 24 species in the region, many rare. Females differentiated from sometimes similar *Tanaecia* (p.89) by stouter bodies.

Malayan Baron ▪ *Euthalia monina* 3.5cm

DESCRIPTION Sexes dissimilar. Males yellow-brown beneath, occurring in forms that differ on the upperside, which may be black with a shining blue or green hindwing marginal band, or brown with pale arrowhead-shaped spots. Female larger, grey-brown with pale helmet-shaped spots on both wings.

Ssp. monina, male-form monina

DISTRIBUTION NE India to S China, through Thailand, Peninsular Malaysia, Singapore, Sumatra, Borneo, Palawan, and Java eastwards to Sumba.
SUBSPECIES Four subspecies: *monina* (Singapore to S Thailand), *insularis* (Tioman Island), *varius* (Langkawi Island) and *kesava* (the rest of Thailand), which is darker, with blue nearly absent.
HABITS AND HABITAT Inhabits primary and disturbed inland forest from the lowlands to highlands. Usually seen along forest tracks.

Ssp. monina, female

Baron ▪ *Euthalia aconthea* 3.5cm

DESCRIPTION Male brown with dark bands or spots on both wings and a few small, white spots near the forewing apex. Underside paler, with dark spots and lines. Female larger and paler, with bigger spots.
DISTRIBUTION Sri Lanka and India to S China, through Thailand, Peninsular Malaysia, Singapore, Sumatra, Borneo, Palawan, and Java eastwards to Sumbawa.

Male of ssp. gurda

SUBSPECIES The ssp. *gurda* occurs from Singapore to S Thailand, and the more contrastingly marked ssp. *garuda* in the rest of Thailand and in Langkawi Island.
HABITS AND HABITAT Commonest in villages and gardens where mango, its host plant, is cultivated, but also occurs in forest in the lowlands.

Female of ssp. gurda

Male of ssp. pinwilli

Green Baron ■ *Euthalia adonia* 3.5cm

DESCRIPTION One of a group of 5 barons that have black and metallic green wings, spotted red and white. Male forewing with small white spots but no red in the cell. Female with large white spots, characteristically continuing to the hindwing. **Whitehead's Green Baron** *Euthalia whiteheadi* (S Thailand and Peninsular Malaysia) is a highland species, and the **Gaudy Baron** *Euthalia lubentina* is commoner in Thailand.

DISTRIBUTION S Thailand, Peninsular Malaysia, Singapore, Sumatra, Borneo, Palawan, Java and the Lesser Sunda Islands.

Female of ssp. pinwilli

SUBSPECIES Two subspecies in the region: *pinwilli* (Peninsular Malaysia) and *beata* (Langkawi Island and S Thailand).

HABITS AND HABITAT Occurs at low to moderate elevations and is never common. More frequently encountered in villages and the countryside than in forest.

Grand Duchess ■ *Euthalia patala* 5.5cm

DESCRIPTION Large. Sexes are alike, brown above, overlaid with greenish scales that shimmer in strong light, and with a band of white spots from the forewing costa to near the tornus. Two smaller costal spots are present beyond the white band and there are black lines in the cell. Underside greenish white with the forewing markings visible. The **French Duke** *Euthalia franciae* is smaller, with the wing spots narrower and running down both wings.

DISTRIBUTION N India to N Thailand and N Laos.

SUBSPECIES Occurs as ssp. *taooana*.

HABITS AND HABITAT Found in forested submontane and montane regions of northern Thailand where it is rare. Flies fast in the understorey, descending to the ground occasionally, and usually landing with wings spread.

Male

COUNTS – TANAECIA

COUNTS – TANAECIA

Thinner-bodied than barons (p.87), with less pointed wing tips and slightly slower flight. There are 12 species in the region.

Malayan Viscount ■ *Tanaecia pelea* 3.5cm

DESCRIPTION Grey-brown above with pale grey to brown arrowhead-shaped discal spots. Underside paler with a purplish tint. Female larger with more elongate forewings. There are 3 similar and smaller species, the smallest being the **Small Viscount** *Tanaecia aruna* (S Thailand and Peninsular Malaysia), which has 2 detached spots inner to the forewing markings.

Female of ssp. pelea (Peninsular Malaysia)

DISTRIBUTION S Burma and S Thailand through Peninsular Malaysia, Singapore, Sumatra, Java and Borneo.

SUBSPECIES Two subspecies: *pelea* from S Thailand to Singapore, and the darker purple-brown ssp. *irenae* on Tioman Island.

HABITS AND HABITAT Flies in the understorey, landing on leaves with wings spread. Widely distributed in forests at low to moderate elevations.

Male of ssp. pelea (Singapore variant)

Common Earl ■ *Tanaecia julii* 3.5cm

DESCRIPTION Male brown above, with a blue marginal hindwing band that is more restricted to the hindwing tornus and termen than in other blue-banded counts. Female all brown above, or with prominent, white apical spots (ssp. *odilina*). Underside yellow-brown, with small black markings and, in the female, overlaid with pale bluish scaling on the hindwing.

DISTRIBUTION N India to S China, Thailand and Peninsular Malaysia.

Male of ssp. bougainvillei

SUBSPECIES Three subspecies: *bougainvillei* (Peninsular Malaysia), *xiphiones* (S and W Thailand, and Langkawi Island) and *odilina* (the rest of Thailand).

HABITS AND HABITAT Flies fast in the understorey and along forest tracks. Locally common in forests in the lowlands and highlands.

Female of ssp. xiphiones

Horsfield's Baron ■ *Tanaecia iapis* 3.5cm

DESCRIPTION Male black-brown above, with a blue marginal hindwing band that extends narrowly along the forewing termen. Underside yellowish brown. The female resembles the Malayan Viscount but with whiter hindwing markings and faintly blue-edged forewing markings. The **Malayan Count** *Tanaecia godartii* (S Thailand and Peninsular Malaysia) is very similar in the male but has a more prominent and straighter underside forewing dark line that is angled instead of parallel to the forewing margin. The female has oblong instead of arrow-like markings on the forewing. Two larger species are the **Blue Count** *Tanaecia flora* (S Thailand and NW Peninsular Malaysia) and

Male

Grey Count *Tanaecia lepidea* (Thailand, except the south, and the highlands of Peninsular Malaysia), which have a pale blue and grey marginal band, respectively.
DISTRIBUTION S Thailand to Java, including Singapore, and from Sumatra to Palawan.
SUBSPECIES Represented by ssp. *puseda*.
HABITS AND HABITAT Similar in behaviour to the Malayan Viscount (p.89). Common in forested areas up to moderate elevations.

Female

Yellow Archduke ■ *Lexias canescens* 3.8cm

DESCRIPTION Both sexes are black above with numerous small yellow spots. Distinguished from females of the Archduke and Dark Archduke (p.91) by their smaller size and yellow instead of blue underside hindwing scaling.
DISTRIBUTION S Thailand, Peninsular Malaysia, Singapore, Sumatra including Nias, and Borneo.

SUBSPECIES A single ssp. *pardalina* occurs in this region.
HABITS AND HABITAT Can be common in some lowland forests, where it flies low and often swiftly. Very fond of feeding on fallen forest fruits.

Archduke ▪ *Lexias pardalis* 4.3cm

DESCRIPTION Stout-bodied. Male black above, with a wide blue marginal hindwing band and a narrower green marginal border on the forewing termen. Blue band black-spotted along its margin. Underside deep rusty brown. Female larger, resembling a large Yellow Archduke, but bluish instead of yellow on the hindwing underside. The almost identical **Dark Archduke** *Lexias dirtea* (Thailand to Singapore) has black-instead of orange-tipped antennae when viewed from above. Both species have the lower antennal surface orange.
DISTRIBUTION NE India to Vietnam and S China, through Thailand, southwards to Singapore and Java, and from Sumatra eastwards to Palawan.
SUBSPECIES The ssp. *dirteana* occurs from Singapore to S Thailand, and ssp. *jadeitina* north of this range.
HABITS AND HABITAT Commoner than the Yellow Archduke, which it resembles in behaviour. Restricted to lowland forest, in which it can sometimes be locally abundant.

Male of ssp. dirteana

Panther ▪ *Neurosigma siva* 3.8cm

DESCRIPTION An unmistakable butterfly with a chequered black, white and orange appearance. Basal half of forewing white, with the base orange, and with black blotches. The rest of both wings are black with white quadrangular, rounded or arrowhead-shaped spots and streaks. The female is darker looking than the male.
DISTRIBUTION NE India to N Thailand, Laos and Vietnam.
SUBSPECIES Represented by the eastern ssp. *nonius* in N Thailand, which has a much reduced orange area on the forewing.
HABITS AND HABITAT Swift-flying. Rare in forested areas at submontane altitudes. Males come to wet spots on the ground or on riverbanks.

Male

> **TABBIES AND RELATIVES – PSEUDERGOLINAE**
> A small subfamily of the Nymphalidae comprising 7 species in 4 genera. Often seen near streams and rivers.

Constable ▪ *Dichorragia nesimachus* 4.5cm

DESCRIPTION Ground colour dark olive-green with a bluish sheen on the upper half of the hindwing. Both wings with black markings and white zigzagging lines on the outer margins, especially at the forewing apex. Underside dark olive-brown, and with more

prominent white forewing markings. Female greener.
DISTRIBUTION N India to S China and Japan, through Southeast Asia and as far east as New Guinea. Absent from Singapore.
SUBSPECIES Two local subspecies: *deiokes* (up to S Thailand) and *nesimachus* (further north). The former is sometimes considered to be the Sumatran ssp. *machates*.
HABITS AND HABITAT An extremely fast-flying and rather rare butterfly, inhabiting inland forest in the lowlands and highlands.

Male of ssp. deiokes

Popinjay ▪ *Stibochiona nicea* 3.2cm

DESCRIPTION Black, peppered above with small white and blue spots and markings on the forewing termen and apex. Hindwing margin white, spotted with black and dusted

with blue on its inner edge, divided by dark-dusted veins, and usually bordered by a blue wavy line. Underside similar, but lacks blue and is more heavily spotted.
DISTRIBUTION N India to S China, through Thailand and Peninsular Malaysia.
SUBSPECIES Generally occurs as ssp. *subucula* in this region, but ssp. *nicea* occurs in the northern Thai province of Nan.
HABITS AND HABITAT Rare, occurring in forest at moderate elevations. Flies moderately fast in the middle storey. Males may return repeatedly to the same perch.

Ssp. subucula

Emperors – Apaturinae
A subfamily of the Nymphalidae that is most diverse in the subtropical belt between the Himalayas and Japan. Several genera occur from Thailand to Singapore.

Elegant Emperor ▪ *Eulaceura osteria* 3.0cm

Male

DESCRIPTION Male black above with a white band running from mid forewing down the entire hindwing. Underside silvery grey edged with brown, and with 2 faint eyespots. Female upperside brown with white markings on the forewing and dark markings near the wing margins. Underside paler.
DISTRIBUTION S China, Laos, Vietnam, S Thailand, Peninsular Malaysia, Singapore, Sumatra, Java, Borneo and Palawan.
SUBSPECIES The local subspecies is *kumana*.
HABITS AND HABITAT Flies fast in the understorey, landing frequently, usually under leaves, with wings closed. Scarce in forests at low to moderate elevations, but males can be abundant in some localities.

LEFT: *Male*. RIGHT: *Female*

Courtesan ▪ *Euripus nyctelius* 2.8cm

Male of ssp. euploeoides

DESCRIPTION Sexes differ greatly. Male with hindwing produced at the tornus and termen. Upperside black with white streaks and spots; underside chocolate brown with wider white markings. Female much larger (4.0cm), occurring in various female forms, for example *isina* and *euploeoides*, which mimic a male or female Magpie Crow (p.54), respectively. The **Painted Courtesan** *Euripus consimilis* (N Thailand) has red basal and tornal hindwing spots in the male, and its female has very wide white streaks.
DISTRIBUTION N India to S China, Thailand southwards to Singapore and Java, and Sumatra eastwards to the Philippines.
SUBSPECIES The Thai subspecies is *nyctelius*, and the southern ssp. *euploeoides*.
HABITS AND HABITAT Flies slowly unless alarmed. Usually seen along forest roads in the lowlands.

LEFT: *Male of ssp.* euploeoides
RIGHT: *Ssp.* euploeoides,
female-form isina

> **TROPICAL BRUSHFOOTS – BIBLIDINAE**
> A subfamily of Nymphalids most diverse in South America, but 2 slender-bodied genera with angled forewings occur in this region.

Common Castor ■ *Ariadne merione* 2.5cm

DESCRIPTION Four species of castors occur in the region. They are orange-brown above with wavy black lines, and dark brown below with darker wavy bands. The darker males have a greyish hindwing costa. The Common Castor has some of the wavy lines paired above, forming darkened bands, and has a small white spot near the forewing apex. The **Angled Castor** *Ariadne ariadne* (Thailand and Peninsular Malaysia) also has the spot, but the blacker wavy lines are unpaired.

DISTRIBUTION India to S China, through Thailand, Peninsular Malaysia, Sumatra, Java, Borneo and the Philippines.
SUBSPECIES Two regional subspecies: *ginosa* (Peninsular Malaysia and southernmost Thailand) and *tapestrina* (the rest of Thailand).
HABITS AND HABITAT Has a slow, flitting flight. Fairly common at the forest edge in the lowlands.

Male of ssp. ginosa

Blue Dandy ■ *Laringa castelnaui* 2.8cm

DESCRIPTION Male blue above and dark brown below with darker bands and fine dark streaks. Female dull yellow-brown above, with dark bands and patches, and overlaid with fine dark streaks. The male of the rarer **Banded Dandy** *Laringa horsfieldii* (Thailand) is black with blue-white discal areas, and the female is more orange-brown.
DISTRIBUTION S Burma, S Thailand, Peninsular Malaysia, Sumatra, Java, Borneo and Palawan.
SUBSPECIES Occurs as ssp. *castelnaui* in this region.
HABITS AND HABITAT Has a similar but faster flight than a castor. Inhabits forested areas up to moderate elevations.

Male

> ## MAPS AND MAPLETS – CYRESTINAE
> A small subfamily of Nymphalids comprising slender butterflies with dark lines on the wings, resembling the contours of a map. They have a flitting flight, and rest with wings squarely spread.

Little Map ■ *Cyrestis themire* 2.3cm

DESCRIPTION Its small size, flat-tipped forewing and pale brown borders separate this species from other maps. Both wings have pale orange-brown lines on the white areas and, like other maps, the hindwing has a short, pointed tail.

DISTRIBUTION Burma, Thailand, Peninsular Malaysia, Sumatra, Borneo, Palawan, Java and the Lesser Sunda Islands.

SUBSPECIES The mainland subspecies is *themire*, and a number of island subspecies occur off the east coast of the Peninsula: *siamensis* (Perhentian Islands), *robinsoni* (Tioman Island) and *pemanggilensis* (islands of Aur and Pemanggil).

HABITS AND HABITAT Inhabits primary forest in the lowlands and highlands. Common in some forest locations, usually near rivers and streams. Often lands under leaves.

Male of ssp. themire

Marbled Map ■ *Cyrestis cocles* 3.2cm

DESCRIPTION Distinguished from other maps by its relatively large size and round-tipped forewing, as well as its ill-defined border and markings on a yellow-white background. The line-like markings are either greyish brown (form *earli*) or pale brownish yellow (form *formosa*).

DISTRIBUTION NE India to S China, through Thailand, Peninsular Malaysia, Borneo and Sulawesi; absent from Sumatra and Java.

Male of ssp. earli, *form* formosa

SUBSPECIES Occurs as ssp. *earli* from Peninsular Malaysia to S Thailand and as ssp. *cocles* in the rest of Thailand.

HABITS AND HABITAT Uncommon and localised in occurrence in primary dipterocarp forest in the lowlands.

Male of ssp. earli, *form* earli

Male

Common Mapwing
■ *Cyrestis maenalis* 3.0cm

DESCRIPTION Forewing tip pointed and wings white with prominent black lines and band-like borders. Tornal areas with black-spotted orange patches. In the somewhat similar **Straight Line Map** *Cyrestis nivea* (W and S Thailand, and Peninsular Malaysia), the black lines are finer and do not expand into bands towards the forewing costal margin.

DISTRIBUTION Peninsular Malaysia, Sumatra, Borneo and the Philippines.

SUBSPECIES Represented by ssp. *martini*.

HABITS AND HABITAT Occurs in forested areas in the highlands and is generally replaced by the Straight Line Map at moderate elevations. Males of both species are fond of settling on wet ground.

Greater Wavy Maplet ■ *Chersonesia rahria* 2.0cm

DESCRIPTION One of 5 small species of maplets in the region, which are brownish orange with dark bands. The Greater Wavy Maplet is characterised by a diffuse, dark discal band and 2 small black apical marks at the forewing tip, which may be joined. It is difficult

Male

to distinguish from the **Wavy Maplet** *Chersonesia intermedia* (Thailand and Peninsular Malaysia), which has a slightly wider band and a single apical mark. The **Rounded Maplet** *Chersonesia peraka* (S Thailand to Singapore) is paler with rounder wings, and has the discal band bounded by black lines.

DISTRIBUTION S Thailand, Peninsular Malaysia and the Southeast Asian islands as far east as Sulawesi, but extinct in Singapore.

SUBSPECIES The regional subspecies is *rahria*.

HABITS AND HABITAT Has a gentle, flitting flight. Fairly common, keeping to the understorey and edges of forest tracks at low to moderate elevations.

> **TRUE BRUSHFOOTS – NYMPHALINAE**
> A large subfamily of Nymphalidae containing familiar butterflies such as the pansies and European commas.

Intricate Jester ■ *Symbrenthia hypatia* 2.3cm

Male

DESCRIPTION Black with fiery orange streaks and bands, superficially resembling a lascar but larger, with a short, pointed hindwing tail and an intricate web of red-brown lines on a yellowish underside. In the **Common Jester** *Symbrenthia lilaea*, the underside markings form a straight, dark line that contrasts against the other web-like markings. The **Spotted Jester** *Symbrenthia hypselis*, has the underside black spotted. Both occur in Thailand and Peninsular Malaysia.

Male

DISTRIBUTION S Burma, Thailand, Peninsular Malaysia, Sumatra, Java, Borneo, and Palawan to Mindanao.
SUBSPECIES A single ssp. *chersonesia* occurs in the region.
HABITS AND HABITAT Fast-flying and restless, but males may settle on the ground. Uncommon in lowland and highland forests.

Blue Admiral ■ *Kaniska canace* 3.2cm

DESCRIPTION Wing margins wavy, deeply concave at the forewing termen, and hindwing with a short tail. Black above, with a blue band running down both wings. Underside densely and finely streaked, forming contrasting areas of black and brown.
DISTRIBUTION Sri Lanka and India to China and Japan, through Thailand, Peninsular Malaysia, Sumatra, Java, Borneo and the Philippines.
SUBSPECIES In Peninsular Malaysia and southernmost Thailand it is represented by ssp. *perakana*, and north of this by ssp. *canace*.

Ssp. canace

HABITS AND HABITAT Inhabits forest and the forest edge. In Peninsular Malaysia it occurs at higher altitudes. Very fast but usually low-flying. Males often use a perch from which they make quick flights.

Ssp. canace

Male of ssp. bolina

Ssp. bolina, *female-form* nerina

Female of ssp. jacintha

Female of ssp. bolina

Great Egg-fly
■ *Hypolimnas bolina* 4.5cm

DESCRIPTION Very variable. Females occur in different forms. Underside brown, with white marginal markings and white discal bands. Upperside of male blue-black with white to bluish-white discal patches on both wings, and small apical spots which, in ssp. *jacintha*, continue as a line of spots down both wings. Female brown above, with varying white and orange patches (ssp. *bolina*), or with white marginal hindwing borders and sometimes bluish discal patches (ssp. *jacintha*).

DISTRIBUTION Madagascar and Mauritius, and from India to southern Japan, through Southeast Asia (including Singapore) and the Pacific as far as Easter Island.

SUBSPECIES Occurs as 2 originally geographically separate sspp. *jacintha* and *bolina* that now coexist and interbreed.

HABITS AND HABITAT Usually flies low and lands with wings open. Common in wayside vegetation, villages, parks and gardens in the lowlands.

Chocolate Soldier ■ *Junonia iphita* 3.0cm

DESCRIPTION Brown above, with diffuse, narrow dark bands. Hindwing margin with a row of indistinct eyespots. Underside darker with broader dark lines. The richer brown **Spotted Chocolate Soldier** *Junonia hedonia*, has larger and more distinct hindwing eyespots. It occurs in Singapore and Johore and has spread to some other parts of the Peninsula.

DISTRIBUTION India to S China, through Thailand, Peninsular Malaysia, Sumatra, Borneo, Palawan, Java and the Lesser Sunda Islands.

SUBSPECIES The subspecies in Thailand is *iphita* and the subspecies in Peninsular Malaysia *horsfieldi* (sometimes considered to be *iphita*).

HABITS AND HABITAT Can be common where it occurs. Generally keeps to low wayside growth especially at the edges of forest in the lowlands. Less likely to come into open grassy areas than other pansies (*Junonia*).

Ssp. iphita

Grey Pansy ■ *Junonia atlites* 3.0cm

DESCRIPTION Grey above, with several dark wavy lines and a series of black-and-orange or grey eyespots running down both wings. Paler below, with markings less distinct. Spots may be greatly reduced in the dry season.

DISTRIBUTION Sri Lanka and India to S China, through Southeast Asia as far east as Sulawesi.

SUBSPECIES Occurs as a single ssp. *atlites* from Thailand to Singapore.

HABITS AND HABITAT Inhabits coastal areas and open country in the lowlands and sometimes highlands. Common in villages, parks and sometimes gardens. Like most pansies, it usually flies near the ground or among low vegetation, with wing flaps and wing glides, and may intermittently spread and close its wings when resting.

Peacock Pansy ■ *Junonia almana* 2.7cm

DESCRIPTION A distinctive species. Rich, orange-brown above with a prominent black and orange hindwing eyespot and a smaller black forewing eyespot. The wings have black, outer lines that form a narrow border, and the forewing has wavy black costal bars. The underside is pale yellow-brown with clear eyespots, a pale, central line on the hindwing, and narrower, dark, wavy lines on the forewing. Females are paler and browner orange than males.

Male of ssp. javana

DISTRIBUTION Sri Lanka and India to S China and S Japan, through Southeast Asia (including Singapore) as far east as Sulawesi.
SUBSPECIES Represented by ssp. *almana* in Continental Thailand and, to the south, by the smaller ssp. *javana* that is sometimes considered the same as *almana*.
HABITS AND HABITAT Similar in habitat and behaviour to the Grey Pansy but commoner in gardens.

Lemon Pansy ■ *Junonia lemonias* 2.8cm

DESCRIPTION Greyish brown above, but blacker and yellow-spotted in the outer two thirds of the forewing. Both wings have marginal yellow lines as well as two almost equal orange and black eyespots. Underside pale brown with marginal eyespots and with wavy dark lines that may be lightly shaded brown or orange-brown between.

DISTRIBUTION India and Sri Lanka to S China, through Thailand and much of Peninsular Malaysia. Also present in the Philippines.
SUBSPECIES Occurs as a single ssp. *lemonias* in this region.
HABITS AND HABITAT Similar to the Grey Pansy in behaviour. Inhabits open country, villages, parks and gardens in the lowlands. In Peninsular Malaysia, it is commoner in the northwest.

Blue Pansy ■ *Junonia orithya* 2.3cm

DESCRIPTION Male black above with yellow markings across the outer half of the forewing and deep blue over most of the hindwing. Wing margins lined yellow, and each wing with two orange and black eyespots, which may be small or obscured. Underside variegated yellow-brown with darker eyespots. Female browner, with larger eyespots, hindwing blue dull or absent, and with two orange costal forewing bars.

DISTRIBUTION Africa (except the north) and S Arabia, Afghanistan, Sri Lanka and India across to S China and Taiwan, through Southeast Asia and reaching Australia.

SUBSPECIES Two local subspecies: *wallacei* (Singapore to S Thailand) and *ocyale* (to the north).

HABITS AND HABITAT Similar in behaviour to the Grey Pansy, but fond of flying low over short grass. Common in lowland coastal scrub and open country, also frequenting gardens, parks and fields.

ABOVE RIGHT: *Male of ssp.* wallacei
RIGHT: *Female of ssp.* wallacei

Yellow Pansy ■ *Junonia hierta* 2.5cm

DESCRIPTION A striking and easily recognised pansy. Male deep black and bright yellow above with a bright blue patch near the hindwing costa. Female less brightly coloured, with small black eyespots on the yellow areas.

DISTRIBUTION Most of Africa, and from India to S China, through Thailand, and occasionally encountered in NW Peninsular Malaysia and Langkawi Island.

SUBSPECIES The local subspecies is *hierta*.

HABITS AND HABITAT Behaviour similar to that of the Grey Pansy. Locally common in Thailand. Inhabits open country in the lowlands and highlands.

Male

Female

Male of ssp. pratipa

Autumn Leaf ■ *Doleschallia bisaltide* 3.5cm

DESCRIPTION Leaf-like in appearance when wings are closed. Underside brown, with a central line and faint eyespots. Male rusty brown above with a wide black forewing apical border and rusty brown costal bar. Female yellower towards the forewing border.
DISTRIBUTION India to S China, Southeast Asia, Australia and the Bismarck Islands, but absent from New Guinea.
SUBSPECIES Originally 2 local subspecies: *pratipa* (Singapore to S Thailand) with a prominently yellow-banded female, and *continentalis* (to the north and in Langkawi Island).
A recent arrival or introduction thought to be a variety of ssp. *bisaltide* has small white spots on the forewing border and lacks white underside spots. It is now commoner than *pratipa* in Peninsular Malaysia and Singapore.
HABITS AND HABITAT Fast-flying. Uncommon in forested areas and near forest in the lowlands and highlands.

Male of ssp. pratipa

Male

Orange Oakleaf ■ *Kallima inachus* 3.7cm

DESCRIPTION Bears even closer resemblance beneath to a dead leaf than the smaller Autumn Leaf. Forewing tips pointed and out-turned. Upperside steely purple-blue, with an orange forewing band and black forewing apex, and 2 tiny white forewing spots. Female paler. The **Scarce Blue Oakleaf** *Kallima knyvetti* (NW Thailand) has a bluish-white instead of orange band.
DISTRIBUTION N India eastwards through Thailand to S China and southern Japan.
SUBSPECIES The local subspecies is *siamensis*.
HABITS AND HABITAT Flies fast at moderate height in the forest understorey.

Male

Malayan Oakleaf
▪ *Kallima limborgii* 3.7cm

DESCRIPTION Similar to the Orange Oakleaf (p.102) but darker blue and orange, and has less pointed and out-turned forewing tips. Female paler.
DISTRIBUTION S Burma, S Thailand, Peninsular Malaysia, Sumatra and Borneo.
SUBSPECIES The local subspecies is *limborgii*.
HABITS AND HABITAT Flies fast at moderate height in lowland and highland forests and at the edges of forest paths, but is rare. Often perches on tree trunks or branches facing downwards, with hindwing tip touching the trunk.

ABOVE AND LEFT:
Male

Wizard ▪ *Rhinopalpa polynice* 3.3cm

DESCRIPTION Wing shape unusual. Forewing termen deeply concave, and hindwing with a pointed tail. Upperside rich orange-brown with wide black outer borders, and black hindwing marginal spots. Underside dark brown, tinged with orange- or red-brown, and with white-outlined, marginal eyespots and several fine, erratic white lines. Female paler.
DISTRIBUTION Burma to Indo-China, through

Southeast Asia as far as Sulawesi, but absent from Singapore.
SUBSPECIES Two local subspecies: *birmana* in N and W Thailand and *eudoxia* in S Thailand and Peninsular Malaysia.
HABITS AND HABITAT Has a quick, nimble and restless flight, but males may settle on wet ground. Uncommon, inhabiting primary forest up to moderate elevations.

Male of ssp. eudoxia

Male of ssp. eudoxia

Male of ssp. albipunctatus

Punchinello ▪ *Zemeros flegyas* 2.1cm

DESCRIPTION Red-brown, chequered with dark brown spots. The northern subspecies has numerous small white spots. Female paler, with rounder wings. The **Malayan Punchinello** *Zemeros emesoides* (S Thailand and Peninsular Malaysia) is more orange-brown with several dark brown bands.
DISTRIBUTION NE India to S China, and through Southeast Asia. Absent from Singapore.
SUBSPECIES Two subspecies in the region: *albipunctatus* (Peninsular Malaysia and southernmost Thailand) and *allica* (Langkawi Island and most of Thailand).
HABITS AND HABITAT Usually seen at the edges of forest tracks. Males may form small territorial groups, flying fast among low trees in the evening. Fairly common in disturbed and primary forest up to the highlands.

Male of ssp. egeon

Orange Punch ▪ *Dodona egeon* 2.4cm

DESCRIPTION One of 6 species of punches in the region. Recognised by its deep orange-brown, dark-dusted wing bases. Outer halves of wings black with orange spots and stripes; underside red-brown with cream to white spots and stripes. Female larger and paler.
DISTRIBUTION Nepal to S China, through Thailand and Peninsular Malaysia.
SUBSPECIES The Peninsular Malaysian subspecies is

Male of ssp. egeon

confluens and the Thai subspecies is *egeon*.
HABITS AND HABITAT Rare in high-elevation forest. Flies very fast. Males perch on low trees on mountain tops and may come to moist ground.

Columbine ■ *Stiboges nymphidia* 2.0cm

DESCRIPTION Wings white with wide black outer borders that also extend along the forewing costa. Marginal black borders bear small white spots. Underside borders are reddish brown. Female larger, with rounder wings.
DISTRIBUTION N India to Vietnam, through Thailand, Peninsular Malaysia, Sumatra and Java.
SUBSPECIES Occurs as sspp. *nymphidia* (S Thailand to Peninsular Malaysia) and *elodinia* (N Thailand), which is sometimes considered to be a separate species.
HABITS AND HABITAT Behaviour similar to other metalmarks, but it rarely flies high and often lands under leaves. Rare and localised in lowland and highland forests.

Female

JUDIES – ABISARA
Eleven species in the region. Most have a small tail. Flight behaviour is typical of metalmarks, but male plum judies (*echerius* group) may form small territorial groups on low trees in the evening, flying fast and defending a perch.

Spotted Plum Judy ■ *Abisara geza* 2.1cm

DESCRIPTION Hindwing with a triangular tail. Reddish brown above; male with a purple wash when viewed at some angles. Underside paler and crossed by greyish bands. Males are characterised on the upperside by a whitish apical forewing bar and 2 black marginal hindwing spots. Female paler with a clear white forewing bar. The male **Malayan Plum Judy** *Abisara saturata*

Male

(Thailand to Singapore) is recognised by the absence of hindwing spots, and the male **Plum Judy** *Abisara echerius* (Thailand) by its dark upperside wing bands.
DISTRIBUTION S Thailand, Peninsular Malaysia, Singapore, Sumatra, Java, Borneo and Palawan.
SUBSPECIES Represented regionally by ssp. *niya*.
HABITS AND HABITAT Locally common in the forested lowlands, especially in secondary forest.

Female

Malayan Tailed Judy ▪ *Abisara savitri* 2.5cm

DESCRIPTION Has a straight white-tipped tail that is basally triangular. Wings pale brown. Forewing with 2 pale, narrow and almost parallel bands. Hindwing with a single band

and 2 black apical spots above. The larger and darker **Tailed Judy** *Abisara neophron* (Thailand and Peninsular Malaysia, in the highlands) has the inner forewing band wider and at an angle to the outer band. The similarly marked **Dark Judy** *Abisara fylla* (Thailand) lacks tails.
DISTRIBUTION S Thailand, Peninsular Malaysia, Singapore, Sumatra, Java and Borneo.
SUBSPECIES Two local subspecies: *savitri* (Peninsular Malaysia and Singapore) and a whiter-banded ssp. *albisticta* (S Thailand and Langkawi Island).
HABITS AND HABITAT Uncommon in lowland forest, including disturbed habitats.

Male of ssp. savitri

Red Harlequin ▪ *Paralaxita telesia* 2.2cm

DESCRIPTION The 3 species of *Paralaxita* are carmine beneath with black and silvery-blue spots. Among the males, the Red Harlequin is recognised by a white forewing spot

on a black upperside, and the **Banded Red Harlequin** *Paralaxita orphna* (S and SE Thailand, and Peninsular Malaysia) by a white forewing band. The male **Malayan Red Harlequin** *Paralaxita damajanti* (S Thailand and Peninsular Malaysia) is plain carmine above. The females are all carmine above, dusted with brown scaling, most extensively in the Banded Red Harlequin and least in the Malayan Red Harlequin.
DISTRIBUTION Thailand, Indo-China, Peninsular Malaysia, Sumatra and Borneo.
SUBSPECIES Two subspecies: *lyclene* (Peninsular Malaysia and southernmost Thailand) and *boulleti* (most of Thailand).
HABITS AND HABITAT Generally low-flying, keeping to heavy forest cover. Uncommon at low to moderate elevations.

Female of ssp. lyclene

Lesser Harlequin ■ *Laxita thuisto* 1.9cm

DESCRIPTION Both sexes orange-brown below, with blue-edged black spots, bluish basal stripes and whitish apical forewing spots. Male almost black above; female orange-brown above with black spots on both wings and white spots near the forewing apex. Female with rounder wings and with the underside white forewing spots more prominent than in the male.

Male of ssp. thuisto

DISTRIBUTION Burma to Indo-China, through Thailand, Peninsular Malaysia, Singapore, Sumatra, Java and Borneo.

SUBSPECIES Represented by 2 subspecies in the region: *thuisto* in Peninsular Malaysia and S Thailand and *ephorus* to the north.

HABITS AND HABITAT Quick in flight and restless, flying in the understorey. Inhabits forests in the lowlands and is usually rare and localised in occurrence.

Female of ssp. thuisto

Harlequin ■ *Taxila haquinus* 2.5cm

DESCRIPTION The single species in the genus is larger than the Lesser Harlequin, with silvery grey and black spots on a dull orange-brown underside. It is also differentiated by the pale band, instead of spots, across the forewing apex. On the upperside, the male is paler than the Lesser Harlequin and shaded orange-brown across the apex, while the female is duller brown with less distinct black spots.

Male of ssp. haquinus

DISTRIBUTION NE India to Laos and Cambodia, through Thailand, Peninsular Malaysia, Singapore, Sumatra, Java, Borneo and Palawan.

SUBSPECIES Three regional subspecies: *haquinus* (S Thailand and Peninsular Malaysia), *fasciata* (W and N Thailand) and *berthae* (E and SE Thailand).

HABITS AND HABITAT Slower in flight than the Lesser Harlequin. Inhabits inland forest up to moderate elevations and can be common where it occurs.

Female of ssp. haquinus

▪ Blues ▪

Malayan Sunbeam ▪ *Curetis santana* 2.1cm

Male

DESCRIPTION **Sunbeams** (subfamily Curetinae) are orange-red or white above with black borders, and silvery white beneath with faint, dark markings. Ten species in the region. The male Malayan Sunbeam has black forewing borders that extend along the dorsum.

Male

In the male **Burmese Sunbeam** *Curetis saronis* (Thailand to Singapore), the border ends at the tornus.
DISTRIBUTION Burma to Indo-China, Thailand, Peninsular Malaysia, Singapore, Sumatra, Java and Borneo.
SUBSPECIES The regional subspecies is *malayica*.
HABITS AND HABITAT Fast-flying in the middle storey. Common but localised, inhabiting forest up to the highlands.

Sumatran Gem ▪ *Poritia sumatrae* 1.8cm

Male

DESCRIPTION The gems (*Poritia*) usually have disjointed or wavy grey or brown bands beneath. The male Sumatran Gem is recognised by its shining blue-green upperside that has the upper two-thirds of the forewing, including the cell, all black. The female is recognised by its light purple upperside with orange and brown marginal markings.
DISTRIBUTION S Burma, S Thailand, Peninsular Malaysia, Singapore, Sumatra and Borneo.
SUBSPECIES Occurs as ssp. *sumatrae*.

Male   Female

HABITS AND HABITAT Uncommon and inconspicuous in lowland dipterocarp forest, but may bask in the sun at the forest edge.

Blue Brilliant ▪ *Simiskina phalia* 1.8cm

DESCRIPTION Brilliants (*Simiskina*) differ from gems in having lines instead of bands beneath, which may be narrow, or edged orange-brown, or wide and white. Unlike related Bluejohns (*Deramas*), the lines are wavy on the hindwing. The male Blue Brilliant is recognised by its purplish-brown underside that is paler near the margins,

Female

and marked with orange-brown lines. Above it is black with blue-green spots. Females are orange above with black forewing borders.
DISTRIBUTION E and S Burma to Indo-China, through Thailand, Peninsular Malaysia, Sumatra, Borneo and Palawan.
SUBSPECIES The regional subspecies is *potina*.
HABITS AND HABITAT Similar in behaviour to the Sumatran Gem. Rare in lowland forest.

Male

> ### BROWNWINGS – MILETINAE
> A subfamily of generally brown, grey or white Lycaenids that prey on specific aphids, treehoppers or mealy bugs as caterpillars, and feed on their honeydew secretions as adults.

Biggs' Brownwing ▪ *Miletus biggsii* 1.8cm

DESCRIPTION Identifiable among the 14 species of brownwings (*Miletus*) in the region by its smaller size and oblique white forewing band on a brown upperside. Pale brown beneath, with many short, bar-like spots. Female larger, with a wider white band. The **Small Brownwing** *Miletus gaesa* (Peninsular Malaysia and northernmost Thailand) has a virtually unmarked upperside.
DISTRIBUTION S Thailand, Peninsular Malaysia, Singapore, Sumatra and its western islands, and Borneo.
SUBSPECIES The subspecies in this region is *biggsii*.
HABITS AND HABITAT Flies with a rapid up-and-down movement. Males may defend a perch on middle-storey trees. Common but localised in forest up to moderate elevations.

Male

Lesser Darkwing ▪ *Allotinus unicolor* 1.5cm

DESCRIPTION Darkwings (*Allotinus*) differ from brownwings (*Miletus*) in having grey-brown spots and fine streaks instead of bars on a greyish instead of brown underside.

There are 17 rather similar species in the region that vary greatly in size. The Lesser Darkwing is intermediate in size and can be distinguished by the hindwing underside, which has the second postdiscal spot from the costa much displaced inwardly.

DISTRIBUTION NW India to Laos, through Thailand, Peninsular Malaysia, Singapore, Sumatra, Java and Borneo to the Philippines and Sulawesi.

SUBSPECIES Two regional subspecies: *unicolor*, from Singapore to S Thailand, and *rekkia* to the north.

HABITS AND HABITAT Behaviour similar to Biggs' Brownwing, but usually flies much lower. The commonest darkwing, inhabiting secondary and primary forest in the lowlands.

Female of ssp. unicolor

Pale Mottle ▪ *Logania marmorata* 1.1cm

DESCRIPTION Mottles (*Logania*) are small, with the underside mottled brown, black and white. The Pale Mottle is one of 5 species. It has a wavy forewing margin and uniformly mottled underside, and the basal half of the forewing above is grey-white. The female has some white on the hindwing.

DISTRIBUTION S and E Burma to Indo-China, through Thailand, Peninsular Malaysia, Singapore, Sumatra, Borneo, the Philippines and Java to the Lesser Sunda Islands.

SUBSPECIES The ssp. *damis* is found in Singapore, the Peninsula and S Thailand, and ssp. *marmorata* to the north of this range.

HABITS AND HABITAT Flies with a more fluttering flight than other brownwings (Miletinae). Uncommon, inhabiting forests and the forest edge in the lowlands.

Male of ssp. damis

Apefly ■ *Spalgis epius* 1.0cm

Male

DESCRIPTION The name apefly derives from the resemblance of the pupa to the face of a monkey. Small, with very short antennae. Upperside brown, with a touch of white on the forewing, which is more pronounced in the female. Underside

Female

grey, with dark, narrow, wavy streaks, and with a small pale forewing spot beyond the cell.
DISTRIBUTION India and Sri Lanka eastwards to S China, through Southeast Asia (including Singapore) to W Irian.
SUBSPECIES The subspecies occurring in the region is *epius*.
HABITS AND HABITAT Stronger in flight than most brownwings. In the caterpillar stage it feeds on mealy bugs. Uncommon, but found in villages and parks as well as forest in the lowlands.

TRUE BLUES – POLYOMMATINAE
A large subfamily of Lycaenids. Many are a shade of shining blue above, or black and white, with short, delicate tails.

Common Pierrot
■ *Castalius rosimon* 1.5cm

DESCRIPTION Hindwing with short, delicate tails. Both wings white above with black quadrangular spots, black borders, and shining pale bluish scaling at the wing bases. Underside white with black spots and a black basal stripe.
DISTRIBUTION India and Sri Lanka to S China, through Thailand, Peninsular Malaysia, Singapore, Sumatra, Borneo, the Philippines and Java to Sulawesi.
SUBSPECIES The single subspecies in the region is *rosimon*.
HABITS AND HABITAT Has an erratic, fluttering flight. Can be common in secondary forest in the lowlands, and in villages and parks where its host plant Chinese Date is grown.

Male

Banded Blue Pierrot
■ *Discolampa ethion* 1.2cm

DESCRIPTION Male blue above with wide black outer borders and a white band from mid-forewing to hindwing dorsum. Female with blue replaced by black. Underside white with a characteristic double black band near the wing bases, and with outer spots and blotches. Wing margins black-spotted.
DISTRIBUTION India and Sri Lanka to S China, Thailand, Peninsular Malaysia, Sumatra, Borneo, the Philippines and Java to Sulawesi.
SUBSPECIES Two regional subspecies: *thalimar* in Peninsular Malaysia and S Thailand, and *ethion* in the rest of Thailand.
HABITS AND HABITAT Quick-flying, usually seen along forest roads. Males may settle on wet ground. A localised lowland forest inhabitant.

Male of ssp. ethion

Elbowed Pierrot ■ *Caleta elna* 1.7cm
DESCRIPTION Upperside black with a white discal band from mid-forewing to hindwing dorsum. Underside white with black outer spots and blotches, and a single black band near

the wing bases, which is bent (elbowed) on the forewing. In the lightly marked **Angled Pierrot** *Caleta decidia* (N, NE and W Thailand), the forewing has a black basal streak, while in the **Straight Pierrot** *Caleta roxus* (Thailand and Peninsular Malaysia), the black forewing band is straight.
DISTRIBUTION N India to S China, through Thailand, Peninsular Malaysia, Singapore, Sumatra, Borneo, Palawan and Java to Sulawesi.
SUBSPECIES Three regional subspecies: *elvira* (Singapore to S Thailand), *noliteia* (to the north), which has browner outer spots on the hindwing and *epeus* (Tioman Island).
HABITS AND HABITAT Behaviour similar to the Banded Blue Pierrot. Common in secondary and primary lowland forest.

Male of ssp. elvira

Inornate Blue ■ *Neopithecops zalmora* 1.2cm

DESCRIPTION Brown above, with some white on the forewing, which is more prominent in females and may be absent in males. Underside white with a prominent black costal hindwing spot and rows of black-brown dashes on the termens of both wings. The similar-looking but unrelated **Forest Cupid** *Pithecops corvus* (Peninsular Malaysia and Thailand) is recognised by 2 black costal dashes on the forewing, and often has orange-brown marginal scaling.
DISTRIBUTION Sri Lanka and India to S China, through Southeast Asia (including Singapore) as far east as Timor, but absent from Sulawesi.
SUBSPECIES Occurs as ssp. *zalmora*.
HABITS AND HABITAT Flies in the understorey with a weak, flapping flight, as does the Forest Cupid. Common where it occurs in lowland forest.

Male

Malayan Pied Blue ■ *Megisba malaya* 1.1cm

DESCRIPTION Very small, with a short tail. Upperside grey-brown with some white, varying from a small, diffuse forewing patch to a bar from mid-forewing to hindwing. Underside greyish white with small brown and black spots and bars, especially along the termens. Hindwing with two prominent black costal spots. Female with rounder wings.
DISTRIBUTION Sri Lanka and India to S China and Taiwan, through Southeast Asia, including Singapore.
SUBSPECIES The local subspecies is *sikkima*. Further north in Thailand, populations tend to have more white on the upperside.
HABITS AND HABITAT Flies at bush height, with a stronger flight than the Inornate Blue. Sometimes common in coastal and inland secondary forest, occurring also in lowland primary forest.

Male

Male of ssp. lambi

Common Hedge Blue
▪ *Acytolepis puspa* 1.6cm

DESCRIPTION One of many similar-looking, tailless blues of the *Lycaenopsis* group. Male shining blue above with black borders and sometimes a whitish discal patch. Female dark brown with white discal areas. Underside greyish white with black spots, including a characteristic small spot inner to 2 large costal hindwing spots. The **Plain Hedge Blue** *Celastrina lavendularis* (Thailand and Peninsular Malaysia) is duller beneath, with white-edged dark markings.

DISTRIBUTION India to S China, Thailand, Peninsular Malaysia, Singapore and the Southeast Asian islands.

SUBSPECIES Three regional subspecies: *lambi* (Singapore to S Thailand), *volumnia* (Islands of Tioman, Aur and Pemanggil) and *gisca* (further north in Thailand), which has larger spots and more white above in the male.

HABITS AND HABITAT Quick-flying. Males may settle on wet sand. Common in villages and forests in the lowlands and highlands.

Male of ssp. lambi

White Hedge Blue
▪ *Udara akasa* 1.3cm

DESCRIPTION Small; underside white with dark spots. Upperside also white, with a broad, black apical forewing border, hindwing marginal spots, and black and blue dusting along the forewing costa and wing bases. Female with a black costal forewing border. The **Pale Hedge Blue** *Udara dilecta* (Thailand and Peninsular Malaysia) is similar beneath, but has a wavy line bordering the marginal spots, and the male is pale blue above.

DISTRIBUTION Sri Lanka, S India, southern Laos and Vietnam, Peninsular Malaysia, Sumatra and Java.

SUBSPECIES Occurs as ssp. *catullus*.

HABITS AND HABITAT Fast-flying. A highland species that is common in open vegetation.

TOP: *Male.* ABOVE: *Female*

Lesser Grass Blue ■ *Zizina otis* 1.2cm

DESCRIPTION Male purple-blue above with dark brown outer borders. Female dark brown above, and purple in the basal half. Underside pale grey-brown with tiny brown spots. Separable from other similar small species by the absence of forewing cell spots and mid-costal spots, and by the hindwing discal spots, which are out of line near the costa. The more prominently spotted **Dark Grass Blue** *Zizeeria karsandra* (Peninsular Malaysia and Thailand) has these spots on an even curve and has a

Ssp. lampa

forewing cell spot.
DISTRIBUTION N India to Japan, and throughout Southeast Asia.
SUBSPECIES Two subspecies: *lampa* (Singapore to S Thailand) and *sangra* (the rest of Thailand).
HABITS AND HABITAT Flies near the ground in open grassy areas. Common in coastal scrub, gardens, parks, villages and roadsides up to moderate elevations.

Male of ssp. lampa

Tiny Grass Blue ■ *Zizula hylax* 0.9cm

DESCRIPTION The smallest butterfly in the region. Resembles the Lesser Grass Blue, but the hindwing discal spots are arranged in an even curve as in the Dark Grass Blue, from which it can be differentiated on the forewing by the absence of a cell spot but the presence of a mid-costal spot. Female almost all brown above.
DISTRIBUTION South Arabia and Africa; India to S China, through Southeast Asia, Australia, New Guinea, the Solomon Islands and Vanuatu.
SUBSPECIES The subspecies occurring from Singapore to S Thailand is *pygmaea*, while ssp. *hylax* occurs in the north.
HABITS AND HABITAT Similar to the Lesser Grass Blue but restricted to the lowlands.

Ssp. pygmaea

Plains Cupid ■ *Chilades pandava* 1.4cm

DESCRIPTION Hindwing delicately tailed. Male blue above with a thin marginal border and hindwing tornal spot. Female paler blue with wide borders, hindwing marginal spots, and with the spot next to the tail orange-crowned. Underside pale brown with small but distinct black spots, and with the 2 spots near the tail prominently crowned orange. The tailless **Lime Blue** *Chilades lajus* (Thailand and NW Peninsular Malaysia) is paler blue, has larger hindwing spots and lacks orange.

DISTRIBUTION India to China, through Thailand, Peninsular Malaysia, Singapore, Sumatra, Java, Borneo and Palawan.
SUBSPECIES The local subspecies is *pandava*.
HABITS AND HABITAT Sometimes common in beach habitats or in gardens and parks in the lowlands wherever its host plants *Cycas* occur.

Male

Female

Silver Forget-me-not ■ *Catochrysops panormus* 1.6cm

DESCRIPTION Male pale silvery blue above with a black tornal spot. Female brown with bluish wing bases and light blue markings. Underside brownish white with short brown bars and an orange-crowned black tornal spot. In the **Forget-me-not** *Catochrysops strabo* (Thailand to Singapore), the tiny costal spot on the forewing underside is well separated from the band, and the male is darker blue. The less related but quite similar **Gram Blue** *Euchrysops cnejus* (Thailand to Singapore) has 2 orange-crowned black spots beneath, and the male is more purple.

Male

DISTRIBUTION India to S China, through Southeast Asia (including Singapore), Australia, New Guinea and the Solomon Islands.
SUBSPECIES Represented by ssp. *exiguus*.
HABITS AND HABITAT Common in open country and forest clearings in the lowlands, usually flying low among flowering shrubs.

Male

Peablue ■ *Lampides boeticus* 1.7cm

DESCRIPTION Underside buff brown with disjointed dark and pale bands, and with the appearance of a whitish outer stripe. Male purple-blue above; forewing termen with diffuse brown borders. Female brown with bluish wing bases and whitish hindwing markings. Hindwing tornus with 2 black spots that are orange-crowned on the underside.

DISTRIBUTION Southwestern Europe eastwards through most of Africa to E Asia, Southeast Asia (including Singapore),

Male

Australia and the Pacific Islands as far as Hawaii.

SUBSPECIES None. It is similar throughout its range.

HABITS AND HABITAT Migratory. Common in villages, parks, plantations and wayside areas from the lowlands to highlands, wherever its host plants from the bean family grow.

Male

Common Cerulean ■ *Jamides celeno* 1.7cm

DESCRIPTION The 17 species of ceruleans in the region are a shade of blue, purple or green above, and grey-brown below, marked with whitish dashes that pair to form bars.

The Common Cerulean is recognised by its very pale blue upperside. The male has a narrow, brown border along the forewing termen that expands slightly at the apex; the female has wide borders. The **Royal Cerulean** *Jamides caeruleus* (Thailand to Singapore) is a brighter cerulean blue than other species.

DISTRIBUTION India to S China, through Southeast Asia, and as far as New Guinea and the Bismarck Islands.

SUBSPECIES Two subspecies: *celeno* (Singapore to S Thailand) and *aelianus* (to the north).

HABITS AND HABITAT Flies erratically along forest edges or paths. Common in or near forests and plantations up to the mid-highlands.

Ssp. celeno

Transparent Six-line Blue ■ *Nacaduba kurava* 1.5cm

DESCRIPTION Blues of the genus *Nacaduba* are often darker above than ceruleans and have browner undersides. The Transparent Six-line Blue is not easily identified among the 15 regional species. The forewing underside has a pair of mid-cell dashes. Males are a slightly grey purple-blue above, with underside markings partially visible. The broadly brown-bordered female is pale shining blue with a tinge of white on the forewing disc, and with black hindwing marginal spots. The smaller **Rounded Six-line Blue** *Nacaduba berenice* (Thailand to Singapore) has rounder wings, and the male is paler above.

Male of ssp. nemana

DISTRIBUTION India to Japan, through Southeast Asia, Australia, New Guinea and the Solomon Islands. Absent from Singapore.
SUBSPECIES Two regional subspecies: *nemana* (Peninsular Malaysia and S Thailand) and *euplea* (north of this range).
HABITS AND HABITAT Fast-flying. Males settle on riverbanks or defend a perch on low trees, especially on hill-tops. Common in forests up to the highlands.

Pointed Line Blue

■ *Ionolyce helicon* 1.3cm

DESCRIPTION The single species of *Ionolyce* in the region resembles *Nacaduba*, but is best recognised by its very pointed forewing and generally smaller size. Male dark blue above. Female brown with a light blue forewing patch.
DISTRIBUTION Sri Lanka and India to Indo-China, Southeast Asia (including Singapore), Australia, New Guinea and New Ireland.
SUBSPECIES The local subspecies is *merguiana*.
HABITS AND HABITAT Fast-flying. Males are most often seen settling on wet forest roads or sandy riverbanks. Common in lowland and highland forests.

Male

Common Line Blue ■ *Prosotas nora* 1.2cm

DESCRIPTION The 9 species of *Prosotas* may be tailed or tailless, and are generally smaller and browner beneath than *Nacaduba*. The Common Line Blue is tailed, dark blue above, and brown below, marked with darker bars. Female brown above. The **Margined Line Blue** *Prosotas pia* (Peninsular Malaysia and Thailand) has a black upperside marginal border in Thailand, while in Peninsular Malaysia the underside

Male of ssp. superdates

forewing marginal markings are faint. The **Tailless Line Blue** *Prosotas dubiosa* (Singapore to Thailand) is grey-brown below and lacks a tail.

DISTRIBUTION India to Indo-China, through Southeast Asia, Australia and New Guinea.

SUBSPECIES Two subspecies in the region: *superdates* (Singapore to S Thailand) and *ardates* (further north).

HABITS AND HABITAT Similar to the Transparent Six-Line Blue.

Female of ssp. superdates

Ciliate Blue ■ *Anthene emolus* size 1.5cm

DESCRIPTION Tailless. Thicker-bodied than most related blues. Male purple blue above; female brown with purple wing bases. Underside dull brown with indistinct bands outlined by white dashes. Hindwing with a small orange-crowned tornal spot. The **Pointed Ciliate Blue** *Anthene lycaenina* has a more angled hindwing tornus and a small dark costal spot on the hindwing underside.

DISTRIBUTION India to S China, Thailand, Peninsular Malaysia, Singapore, Sumatra, Borneo, Palawan, and Java eastward to Sumbawa.

Male of ssp. goberus

SUBSPECIES Two regional subspecies: *goberus* (Singapore to S Thailand) and *emolus* (to the north).

HABITS AND HABITAT Fast-flying. Males settle on wet ground. The commonest of 3 species of Ciliate Blues (*Anthene*) in the region, inhabiting villages, parks and forest in the lowlands and highlands.

Male of ssp. goberus

Purple Sapphire ■ *Heliophorus epicles* 1.5cm

DESCRIPTION The sapphires (*Heliophorus*) alone represent the **Coppers** subfamily (Lycaeninae) in this region, with 6 species. The tailed Purple Sapphire is deep yellow below with a bright red marginal band on both wings, edged white inwardly. Male black-brown above, with a basal purple forewing patch, an orange forewing bar and orange wing margins. Female with wider orange markings above, and without purple.

DISTRIBUTION N India to SE China, through Thailand, Peninsular Malaysia, Sumatra and Java.

Ssp. tweediei

SUBSPECIES The regional subspecies are *tweediei* (Peninsular Malaysia) and *latilimbata* (Thailand).

HABITS AND HABITAT Fast, but low-flying, preferring open sunny areas with shrubs near forest, where it often basks with wings open. Locally common in the highlands.

FAR LEFT: *Male of ssp.* tweediei
LEFT: *Female of ssp.* tweediei

Long-banded Silverline
■ *Spindasis lohita* 1.6cm

DESCRIPTION The **Silvers** subfamily (Aphnaeinae) is represented by the silverlines (*Spindasis*) in this region, with 12 species, which have a double tail. They are yellow to white below, with silver-centred dark stripes and an orange tornus. The Long-banded Silverline can be recognised by its red-brown stripes and small size. The equally common and similarly sized **Club Silverline** *Spindasis syama* (p.14) is found from Thailand to Singapore. It has black-brown stripes, and the basal

Ssp. senama

forewing streak never reaches the first forewing stripe.

DISTRIBUTION India to S China, through Thailand, Peninsular Malaysia, Singapore, Sumatra, Java, Borneo and Palawan.

SUBSPECIES Two local subspecies: *senama* (Singapore to S Thailand) and *himalayanus* (in the north).

HABITS AND HABITAT Very fast-flying, keeping to bush height and resting frequently. Common in bushland and secondary forest up to the highlands.

Male of ssp. senama

Centaur Oakblue ▪ *Arhopala centaurus* 2.8cm

DESCRIPTION Recognised by its large size and silvery green edging on the underside forewing cell spots. Underside brown markings contrasting. Upperside of male shining purple-blue; female paler purple.

DISTRIBUTION India to S China, Thailand, Peninsular Malaysia, Singapore, Sumatra, Borneo and the Philippines.

SUBSPECIES The local mainland subspecies is *nakula* and the subspecies on Tioman Island is *dixoni*.

HABITS AND HABITAT The only oakblue that occurs in villages, parks and gardens as well as lowland forest. A strong flier, usually keeping to canopies of small trees. Its larvae are looked after by the tailor ant *Oecophylla smaragdina*.

Ssp. nakula

White-dotted Oakblue ▪ *Arhopala democritus* 1.8cm

DESCRIPTION Identified by its short tails, bright blue upperside and irregular, pale underside markings. Female paler blue.

DISTRIBUTION Central Burma, Thailand and Peninsular Malaysia.

SUBSPECIES The ssp. *democritus*, occurring in Thailand and NW Peninsular Malaysia, is paler and more shining above with clearer white markings beneath compared to the main peninsular ssp. *lycaenaria*, which also occurs in southernmost Thailand and may be a distinct species.

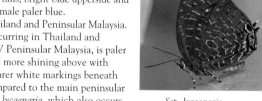

Ssp. lycaenaria

HABITS AND HABITAT Inhabits dipterocarp forests in the lowlands. Flies for short distances in the understorey, landing frequently.

Ssp. democritus

Shining Plushblue ■ *Flos fulgida* 1.8cm

DESCRIPTION Plushblues differ from oakblues in having wider, continuous underside banding. Eight tailed or tailless species in the region. The Shining Plushblue is tailed

and characterised by contrasting underside markings and a dark more continuous subbasal hindwing band. Male violet-blue above; female paler with brown borders. In the **Plain Plushblue** *Flos apidanus* (Thailand to Singapore), the underside markings contrast poorly and the wing bases are reddened.

DISTRIBUTION NE India to Indo-China, and from Thailand south to Java, and Sumatra east to the Philippines.

SUBSPECIES Two subspecies in the region: *singhapura* (Singapore to S Thailand) and *fulgida* (to the north).

HABITS AND HABITAT Fast-flying.

Ssp. singhapura

Male of ssp. singhapura

Uncommon in forests up to moderate elevations. Like all plushblues, it prefers more open habitats than oakblues.

White Imperial ■ *Neomyrina nivea* 2.6cm

DESCRIPTION Unmistakable. Recognised by its large size, predominantly white colour and long, filamentous tails. Underside with irregular, darkly outlined greyish bands. Upperside forewing with a border, which is largely shining blue in the male and black

in the female, arching from mid-costa to tornus. Male with additional blue scales on the upper body, wing bases and costa; female with black borders broader and continued narrowly along the hindwing termen.

DISTRIBUTION S Burma, Thailand, Peninsular Malaysia and Sumatra.

SUBSPECIES The ssp. *periculosa* occurs in Peninsular Malaysia and *hiemalis* in Langkawi Island and Thailand.

HABITS AND HABITAT Fast-flying, usually keeping to the canopies of low trees. Often lands under leaves. Uncommon, occurring in dipterocarp forests up to moderate elevations.

Ssp. periculosa

Burmese Acacia Blue
■ *Surendra vivarna* 1.7cm

DESCRIPTION Four species of acacia blues (*Surendra*) occur in the region. The Burmese Acacia Blue can be identified by its regularly shaped hindwing tornus bearing a short tail and tail-stump. Male deep purple-blue above with broad black apical borders. Female steely greyish blue above with diffuse marginal borders. Underside brown with irregular dark lines and green hindwing tornal scales. The **Common**

Acacia Blue *Surendra quercetorum* (Thailand), which has an all-brown female, is recognised by the concavity between its hindwing tornal lobe and tail.
DISTRIBUTION S Burma, Thailand, Laos, Vietnam, Peninsular Malaysia, Singapore, Sumatra, Java, Borneo and Palawan.
SUBSPECIES Occurs as ssp. *amisena* in this region.
HABITS AND HABITAT Fast-flying. Common in secondary forest and wayside vegetation up to moderate elevations, and wherever its host plant Albizia (*Paraserianthes falcataria*) is grown.

Male

Scarce Silverstreak Blue
■ *Iraota rochana* 1.9cm

DESCRIPTION Double-tailed, with an additional tail-stump. Male metallic greenish blue above with a broad black forewing border and narrow black hindwing border. Female brown above. Underside a mixture of pale brown and dark rusty brown with small black tornal markings and prominent white stripes and spots, including a large white stripe below the hindwing costa.
DISTRIBUTION NE India to Indo-China, Thailand, Peninsular Malaysia, Singapore, Sumatra, Borneo, the Philippines and Java to Sulawesi.
SUBSPECIES The local regional subspecies is *boswelliana*.
HABITS AND HABITAT Fast-flying in the canopies of low trees. Uncommon, but occurs in gardens, parks and forests from the lowlands to highlands.

Male *Female*

Greater Tinsel ■ *Catapaecilma major* 1.8cm

DESCRIPTION Tinsels (*Catapaecilma*) have 3 short tails, are purple to blue above, and yellow or brown beneath, with delicate brown and silver streaks. Five species occur in the region. The Greater Tinsel is light brown beneath with rusty-brown, silver and black streaks. It is differentiated from the similar but smaller and rarer **Lesser Tinsel** *Catapaecilma elegans* by an additional brown and silver streak just beyond the forewing cell-end streak.

Male of ssp. emas

DISTRIBUTION India to S China, Thailand, Peninsular Malaysia, Singapore, Sumatra, Java and Borneo.

SUBSPECIES Two subspecies in the region: *emas* (Singapore to S Thailand) and *albicans* (the rest of Thailand).

HABITS AND HABITAT Flies fast among bushes. Generally rare, but occurs in villages and orchards as well as forest in the lowlands.

Yamfly ■ *Loxura atymnus* 1.7cm

DESCRIPTION Forewing square-tipped, and hindwing pointed, extended as a filamentous tail. Wings orange-red above, with a black forewing border along the termen and across the apex. Female with hindwing tornus dark-dusted. Underside paler, with a faint band across both wings. Tail blackened towards its end, and white tipped. The less common **Malayan Yamfly** *Loxura cassiopeia* (southernmost Thailand, and Peninsular Malaysia) is similar but dark-dusted above, especially on the forewing costa.

DISTRIBUTION India to S China, through Southeast Asia as far as Sulawesi.

SUBSPECIES The local regional subspecies are *fuconius* (Singapore to S Thailand) and *continentalis* (to the north).

HABITS AND HABITAT Flies among low bushes, resting frequently. Common in the lowlands in secondary forest and at the forest edge.

ABOVE LEFT: *Ssp.* fuconius
LEFT: *Male of ssp.* fuconius

Horsfield's Branded Yamfly ■ *Yasoda pita* 1.7cm

DESCRIPTION Similar to the Yamfly in its general shape and orange and black colouration, but the forewing termen bulges slightly towards the tornus, and the borders on the upperside extend along the termen of the hindwing. The underside also has more distinct and rounder markings. The female is more heavily dark-dusted and has a dark brown hindwing discal stripe perpendicular to the cell. The **Branded Yamfly** *Yasoda tripunctata* (Thailand) is paler orange-yellow and has a black hindwing stripe in both sexes.

DISTRIBUTION S Burma, S Thailand, Peninsular Malaysia, Sumatra, Java and Borneo.

SUBSPECIES The sole subspecies in the region is *dohertyi*.

HABITS AND HABITAT Similar to the Yamfly in behaviour. Common but localised in occurrence at the edges of submontane and montane forests.

Branded Imperial

■ *Eooxylides tharis* 1.7cm

DESCRIPTION Hindwing with 3 white tails, the central tail very long. Underside rich brownish orange with a white tornal area that streaks outward on its inner margin and is edged by curved black dashes. Prominent black spots are present on the white hindwing tornus. Upperside black-brown; hindwing tornus white with black marginal markings. Male with bluish-white scaling on the basal half of the upperside.

DISTRIBUTION Burma, S Thailand, Peninsular Malaysia, Singapore, Sumatra, Java, Borneo and Palawan.

SUBSPECIES Represented in the region by ssp. *distanti*.

HABITS AND HABITAT Flies with an up-and-down movement in the understorey and along forest roads and paths. Common where it occurs in forested areas in the lowlands and highlands, especially near streams.

Male

Ssp. frigga

Common Imperial
■ *Cheritra freja* 2.1cm

DESCRIPTION Hindwing with 2 white-lined tails, the upper tail very long, the lower short. Male shining purple above, and female brown, both with a black-spotted white hindwing tornal patch. Underside white, turning orange-brown at the forewing apex and costa. Termen with narrow dashes, becoming broader and blacker near the hindwing tornus, which is marked black and blue.

DISTRIBUTION India to Indo-China, Thailand, Peninsular Malaysia, Singapore, Sumatra, Java and Borneo.
SUBSPECIES Three subspecies: *frigga* (Peninsular Malaysia and Singapore), *freja* (Langkawi Island and S Thailand) and *evansi* (Continental Thailand).
HABITS AND HABITAT Fast-flying, usually in the canopies of low trees. Common in the lowlands in primary and secondary forest, and sometimes in parks and villages.

Male of ssp. frigga

Female of ssp. moorei

Common Posy
■ *Drupadia ravindra* 1.6cm

DESCRIPTION Hindwing with 3 white tails, the central tail elongated. Male black above, with an orange forewing patch in ssp. *boisduvalii*, and with the hindwing mostly shining blue. Female dark brown above, with mid-forewing disc or veins orange and hindwing tornus grey. Forewing underside deep orange, with a dark double line; hindwing white, chequered with black bars. Most of the 8 species of posies (*Drupadia*) have chequered hindwing undersides, but a few are more similar to the Common Imperial.
DISTRIBUTION Burma to Indo-China, Thailand south to Java, and Sumatra eastwards to the Philippines.
SUBSPECIES Three subspecies: *moorei* (Singapore to S Thailand), *caerulea* (Tioman Island) and *boisduvalii* (Thailand, except the south).
HABITS AND HABITAT Keeps to bush height, and is not a fast flier. Common up to montane elevations in disturbed and primary forest.

LEFT: *Male of ssp.* moorei
RIGHT: *Male of ssp.* boisduvalii

Dark Posy ■ *Drupadia theda* 1.8cm

DESCRIPTION Underside like the Common Posy, but outer hindwing markings composed of double black lines instead of solid black bars. Male purple above, with orange mid-forewing veins. Female resembles the Common Posy above, but northern subspecies have a big orange forewing patch. The smaller **Blue Posy** *Drupadia scaeva* is mostly white beneath with less heavy markings. The male has a blue upperside patch on both forewing and hindwing.

Female of ssp. thesmia

DISTRIBUTION S Burma to Indo-China, Thailand, Peninsular Malaysia, Singapore, Sumatra, Borneo, the Philippines and Sulawesi.
SUBSPECIES Three subspecies: *thesmia* (Singapore, Peninsular Malaysia and the southeast extreme of Thailand), *renonga* (Langkawi Island and most of S Thailand) and *fabricii* (the rest of Thailand).
HABITS AND HABITAT Similar to the Common Posy, but stronger in flight.

Male of ssp. thesmia

Yellow Onyx ■ *Horaga syrinx* 1.5cm

DESCRIPTION Sometimes called the Ambon Onyx. One of 6 species of rare onyxes in the region. All have 3 delicate tails. The Yellow Onyx is recognised by its clear light blue colour above. Upperside broadly bordered black, with a white forewing cell-end spot of variable size. Female paler with a clearer spot. Underside brownish yellow with a white transverse line that is thick on the forewing and narrower in the lower half of

the hindwing.
DISTRIBUTION NE India to Indo-China, through Southeast Asia to New Guinea.

Ssp. maenala

Male of ssp. maenala

SUBSPECIES The local subspecies are *maenala* (Singapore and Peninsular Malaysia) and *moulmeina* (Thailand).
HABITS AND HABITAT Rare; usually encountered in secondary growth and on hill-tops. Flies fast and low, but settles often.

Male

Burmese Tufted Royal

■ *Dacalana burmana* 1.9cm

DESCRIPTION Hindwing double-tailed. Shining sky blue above with a black apical forewing border. Male with a forewing brand; female paler with a whitish cell-end bar. Underside white-banded. Identified among 6 local species of Tufted Royals (*Dacalana*) by a broken dark underside line and a small orange tornal crown that does not touch the inner line. The **Malayan Tufted Royal** *Dacalana vidura* (S Thailand and all but NW Peninsular Malaysia) has a clear orange tornal patch that touches but does not displace the continuous dark inner line.

DISTRIBUTION S Burma, Thailand, Central Vietnam and NW Peninsular Malaysia.

SUBSPECIES No subspecies occur.

HABITS AND HABITAT Makes quick but short flights in the forest understorey, landing under leaves. Uncommon and local, occurring up to mid-elevations.

Male

Peacock Royal ■ *Tajuria cippus* 2.2cm

DESCRIPTION Double-tailed. Male deep shining blue above with the forewing apical half and hindwing apical margin black. Female paler, greyer blue with dark outer hindwing spots. Underside buff-grey, with an orange-crowned black tornal hindwing spot. Recognised among 19 local species of *Tajuria* by its prominent, broken black underside line. The **White Royal** *Pratapa deva* (Thailand to Singapore) looks like a very small Peacock Royal with subdued underside markings.

Female of ssp. maxentius

DISTRIBUTION India to S China, Thailand southwards to Java, and Sumatra eastwards to Borneo.

SUBSPECIES The ssp. *cippus* occurs in Thailand, and *maxentius* in Peninsular Malaysia and Singapore.

HABITS AND HABITAT Fast-flying in low trees. Common, but not easily seen, inhabiting parks, villages and forests up to the highlands.

Male of ssp. maxentius

Grand Imperial ■ *Neocheritra amrita* 2.3cm

DESCRIPTION Double-tailed, with the lower tail very long. Male shining blue above with the forewing outer two-thirds and hindwing apex black. Female brown above with a white tornal area marked with black spots. Underside mostly orange, with the lower half of the hindwing white, marked with black at the tornus. Distinguished on the underside from several similar species by its strong orange colour and the presence of only 3 tornal black streaks inner to the marginal spots.

DISTRIBUTION S Burma and S Thailand, Peninsular Malaysia, Singapore, Sumatra and Borneo.

SUBSPECIES The local subspecies is *amrita*.

HABITS AND HABITAT Flies moderately fast, usually keeping to the canopies of middle-storey trees, and more often seen on submontane hill-tops and ridges. Rare in forest at all elevations.

Chocolate Royal ■ *Remelana jangala* 1.8cm

DESCRIPTION Hindwing with 2 short, equal-length tails. Upperside dark brown with the basal half dark, shining purple. Female paler with more extensive purple. Underside chocolate brown with a narrow dark line running down both wings. Hindwing tornus with 2 black spots and shining green scaling.

DISTRIBUTION Central India to S China, through Thailand, Peninsular Malaysia, Singapore, and the

Male of ssp. travana

Southeast Asian islands as far east as the Philippines, Timor and Sulawesi.

SUBSPECIES The ssp. *ravata* is found through most of Thailand, and *travana* occurs from S Thailand to Singapore.

HABITS AND HABITAT Very fast-flying and restless butterfly, keeping to bush height. Common in secondary forests and at the forest edge up to moderate elevations.

Male of ssp. ravata

Female of ssp. teatus

Common Tit ■ *Hypolycaena erylus* 1.7cm

DESCRIPTION Hindwing double-tailed. Male dark purple-blue above with a black forewing patch. Female brown above with an indistinct dark band and 2 black tornal spots, which may be diffusely white-edged. Underside grey, browner near the outer wing margins, with cell-end bars and a narrow orange-brown band running down both wings. Hindwing tornal margin with a silvery patch and 2 black spots, the upper spot orange-crowned.

DISTRIBUTION NE India to Indo-China, throughout Southeast Asia, and as far as New Guinea.
SUBSPECIES Occurs as ssp. *teatus* (Singapore to S Thailand) and *himavantus* (north of this range).
HABITS AND HABITAT Fast-flying among bushes and low trees. Common in mangroves, wayside vegetation and secondary forest in the lowlands.

Male of ssp. teatus

Fluffy Tit ■ *Zeltus amasa* 1.5cm

DESCRIPTION Small, with 2 white filamentous tails, the lower tail long. Male upperside black, with bluish-white scales at forewing base and on most of hindwing. Female brown, with hindwing tornus irregularly whitened. Both sexes with 2 black tornal hindwing spots above and below. Underside white, with apical half of forewing and apex of hindwing orange. Both wings with cell-end bars and a narrow orange-brown band. Hindwing costa with a prominent black spot near the base.
DISTRIBUTION India to S China, Thailand southwards to Java, and Sumatra eastwards to the Philippines.
SUBSPECIES The local subspecies are *maximinianus* (Singapore to S Thailand) and *amasa* (the rest of Thailand).
HABITS AND HABITAT Flies moderately fast, usually at bush height at the forest edge. Common in lowland forests.

ABOVE LEFT AND LEFT: *Male of ssp.* maximinianus

Cornelian ▪ *Deudorix epijarbas* 1.9cm

DESCRIPTION One of 5 local species of cornelians (*Deudorix*). Hindwing tail short. Male red above; forewing widely black bordered; hindwing costa broadly black. Female brown above. Underside grey-brown with paired white lines forming bars and bands. Hindwing tornus edged silvery blue beneath, with a black lobe and orange-ringed black spot.
DISTRIBUTION India to S China, through Southeast Asia,

Male of ssp. cinnabarus

New Guinea, Australia and the Pacific Islands as far as Samoa.
SUBSPECIES Two regional subspecies: *cinnabarus* (Singapore to S Thailand) and *amatius* (to the north). Both are sometimes considered to be ssp. *epijarbus*.
HABITS AND HABITAT Very fast-flying, among bushes and low trees. Sometimes common in secondary forest up to the highlands.

Male of ssp. cinnabarus

FLASHES – RAPALA

Fast-flying. Usually red, brown or blue above, and grey, brown or yellow beneath. Hindwing tornus with a single short tail and small lobe. Eighteen local species. Red species resemble cornelians, but have narrower underside bands.

Copper Flash ▪ *Rapala pheretima* 1.9cm

DESCRIPTION Male brown above, with coppery discs; female a subdued, steely blue. Underside brown with darker brown cell-end bars, a forewing cell spot (sometimes absent in the female), and narrow, dark wing bands. Hindwing tornus with a small black lobe and spot, and silvery green scales.
DISTRIBUTION N India to Indo-China, Thailand, Peninsular Malaysia, Singapore, Sumatra, Java and Borneo.
SUBSPECIES Three local subspecies: *sequeira* (Singapore to S Thailand), *petosiris* (the rest of Thailand) and *tiomana* (Tioman Island).
HABITS AND HABITAT Common in bushy secondary growth and the forest edge up to moderate elevations.

Female of ssp. sequeira

Male

Scarlet Flash ■ *Rapala dieneces* 1.6cm

DESCRIPTION Male red and black above, and female brown, both resembling the respective sexes of the Cornelian (p.131), but smaller in size, and brighter red in the male. Underside brownish yellow with a thin, brown line running down both wings, and indistinct cell-end bars. Hindwing tornus scaled silvery green, with a black tornal lobe and an orange-crowned black tornal spot. Can be distinguished from similar species by its less yellow underside and, in the male, the absence of a purple glint above.

DISTRIBUTION N India to Indo-China, Thailand, Peninsular Malaysia, Singapore, Sumatra, Java, Borneo and the Philippines.
SUBSPECIES The ssp. *dieneces* occurs throughout the region.
HABITS AND HABITAT Flies at bush height. Not uncommon in open areas in lowland forest.

Male

Male

Common Red Flash ■ *Rapala iarbus* 1.7cm

DESCRIPTION A combination of characters make this species recognisable. Male orange-red above, with narrower black forewing borders than most flashes. Forewing disc with the veins blackened. Female coppery brown above. Underside grey, with indistinct cell-end bars and a lightly darkened band across both wings, outwardly lined with white. Hindwing tornus with a black lobe, an orange-crowned black marginal spot, and silvery-blue scaling.

DISTRIBUTION India to Indo-China, Thailand, Peninsular Malaysia, Singapore, Sumatra, Borneo, Palawan, Java and the Lesser Sunda Islands.
SUBSPECIES The local subspecies is *iarbus*.
HABITS AND HABITAT Flies among bushes and the canopies of low trees. Common in the lowlands in secondary growth and at the forest edge, and sometimes in villages and parks.

Male

Chequered Flame ■ *Araotes lapithis* 1.4cm

DESCRIPTION Hindwing single-tailed and lobed. Underside forewing distinctive: orange with a black-edged white band. Hindwing underside mostly white, chequered with black spots and bars. Male black above with forewing base and nearly whole hindwing deep blue. Female brown above with whitened hindwing tornus. The **Spotted Spark** *Sinthusa malika* (Thailand and Peninsular Malaysia) is similar in size, shape and upperside colour, but is greyish white beneath, with the forewing apex brown and with narrower black markings.

DISTRIBUTION NE India, Burma, Thailand, Peninsular Malaysia, Sumatra, Java, Borneo and Palawan.

SUBSPECIES Two subspecies: *uruwela* (Peninsular Malaysia to S Thailand) and *lapithis* (north of this range).

HABITS AND HABITAT Moderately quick in flight, keeping fairly low. Rare in lowland forests.

Female of ssp. uruwela

Plush ■ *Sithon nedymond* 1.6cm

DESCRIPTION Has a short hindwing tail and lobe. Sexes differ starkly. Male dark blue above with a black forewing border and black basal scaling that is extensive on the hindwing. White beneath with a wide, chocolate band bordered by brown wing margins. Hindwing tornus marked in black, orange and silvery blue. Female with rounder wings. Brown above with hindwing tornus white; mainly orange beneath with hindwing dorsum whitened, and with more prominent black hindwing tornal markings.

DISTRIBUTION S Burma, S Thailand, Peninsular Malaysia, Sumatra, Java and Borneo.

SUBSPECIES Two regional subspecies: *nedymond* (Peninsular Malaysia and extreme southeast of Peninsular Thailand) and *ismarus* (Langkawi Island and the rest of S Thailand).

HABITS AND HABITAT Keeps to the understorey, making short, quick flights. Uncommon in lowland forest.

ABOVE RIGHT: *Male of ssp.* nedymond
RIGHT: *Female of ssp.* nedymond

SKIPPERS – HESPERIIDAE
A family that may be mistaken for moths. Antennae usually end in a tapering and bent tip after the antennal club. Many have pointed or angular wings, and are fast and difficult to see in flight.

AWLS – COELIADINAE
A subfamily of thick-bodied skippers. Sometimes brightly coloured. They rest with wings closed.

Male of ssp. consobrina

Orange-striped Awlet ▪ *Burara harisa* 2.5cm

DESCRIPTION One of 11 locally occurring awlets (*Burara*). Wings densely streaked orange-red beneath. Lower body, legs and hindwing tornal margin orange-red. Male yellowish brown above with costa orange; female steely blue. The **Great Orange Awlet** *Burara etelka* (Thailand to Singapore) is similar but much larger.
DISTRIBUTION N India to S China, through Southeast Asia to Sulawesi.
SUBSPECIES Two subspecies: *consobrina* (Singapore to S Thailand) and *harisa* (further north).
HABITS AND HABITAT Fast-flying, usually at dawn and dusk, settling under leaves. Uncommon in forests up to moderate elevations.

Male of ssp. chuza

Yellow Banded Awl ▪ *Hasora schoenherr* 2.5cm

DESCRIPTION A distinctive awl (*Hasora*) with a yellow hindwing band. Eyes black. Wings brown above, with yellowish basal scales. Forewing with whitish spots forming 2 short bands. Underside paler, but dark beyond the yellow band, with hindwing tornal lobe blackened.
DISTRIBUTION NE India to Indo-China, Thailand to Java, and Sumatra to the Philippines.
SUBSPECIES Represented by ssp. *chuza* (Singapore to S Thailand) and *gaspa* (to the north).
HABITS AND HABITAT Like all awls, it flies fast, landing briefly, often under leaves. May visit buildings near forest in the early morning. Moderately common in forests up to the highlands.

Common Awl ■ *Hasora badra* 2.3cm

DESCRIPTION Dark brown, and characterised on the hindwing underside by a small white cell spot and black tornal lobe. Underside shades of brown, more contrasting in female, and with a white dash above the tornus. Female with large yellowish white forewing spots. The rarer **Large-spotted Awl** *Hasora quadripunctata* (Peninsular Malaysia) is recognised by its much larger hindwing cell spot. The **Plain Awl** *Hasora mus* (highlands of Peninsular Malaysia) is uniformly yellow-brown with the hindwing tornus edged black.

DISTRIBUTION India to Japan, through Southeast Asia (including Singapore), up to Sulawesi.

SUBSPECIES A single ssp. *badra* occurs in the region.

HABITS AND HABITAT Similar to the Yellow Banded Awl in behaviour. Sometimes common in disturbed or primary forest up to the highlands.

Male

Plain Banded Awl ■ *Hasora vitta* 2.3cm

DESCRIPTION Underside brown with a moderately wide, outwardly diffuse white hindwing band, and a bluish-green basal glaze. Hindwing tornus blackened. Upperside dark brown, with a single small yellowish-white forewing spot in the male, and with further large spots in the female.

DISTRIBUTION India to China, through Southeast Asia and as far as Irian Jaya (western New Guinea).

SUBSPECIES The ssp. *vitta*, which ranges from Singapore to S and Central Thailand, appears to intergrade with *indica*, which occurs in most of Thailand.

HABITS AND HABITAT Similar to the Yellow Banded Awl in behaviour. The commonest of 19 species of awls (*Hasora*) in the region, occurring in forest and at the forest edge in both the lowlands and highlands.

Male of ssp. vitta

Caudate Awl King ■ *Choaspes subcaudatus* 2.7cm

DESCRIPTION Awl kings (*Choaspes*) are metallic green beneath with darkened veins, and with the hindwing tornus orange-yellow marked in black. The black upperside is clothed basally with metallic green scales and hairs, and the hindwing tornus is edged orange. Five similar-looking species occur in the region. The Caudate Awl King is best recognised in the field by its longer tail.

DISTRIBUTION S Burma to Indo-China, through Thailand, Peninsular Malaysia, Sumatra, Java, Borneo and Tawi-Tawi (Philippines).

SUBSPECIES A single ssp. *crawfurdi* is present in the region.

HABITS AND HABITAT Flies fairly fast and low, settling under leaves. Conspicuous by virtue of its colour. Rare and confined to inland forest in the lowlands.

FLATS – PYRGINAE
A subfamily of skippers that have a distinct behaviour of landing with their wings spread flat.

White-banded Flat
■ *Celaenorrhinus asmara* 2.0cm

DESCRIPTION Sexes similar. Dark brown with indistinct black dappling. Forewing with a short white band comprised of 3 angular spots, the middle spot protruding beyond the others. Also has a series of 3 tiny spots nearer the forewing apex. Upperside overlaid with greenish ochreous scales and hairs.

DISTRIBUTION NE India to S China, through Thailand, Peninsular Malaysia, Sumatra, Java, Borneo, Palawan and Sulawesi.

SUBSPECIES The ssp.

Male of ssp. asmara

asmara occurs in Peninsular Malaysia and S Thailand, while *consertus* is found in the rest of Thailand.

HABITS AND HABITAT Fast but low-flying, landing on or under leaves, and sometimes seen visiting flowers. Uncommon, occurring in lowland forest.

Ssp. asmara

Velvet Flat ■ *Celaenorrhinus ficulnea* 2.0cm

DESCRIPTION Sexes similar. Velvet black above, browner beneath. Forewing with an almost oblong, white band that does not reach the wing margins. Palpi edged orange. The much rarer **Orange-banded Flat** *Celaenorrhinus ladana* (southernmost Thailand and Peninsular Malaysia) has a large orange band. In the commoner **Dark Yellow-banded Flat** *Celaenorrhinus aurivittatus* (Thailand and Peninsular Malaysia) the band is narrower and often yellower, and there is a small orange-yellow bar near the apex. Seventeen species of *Celaenorrhinus* occur in the region.

DISTRIBUTION NE India to S Burma and S Thailand, Peninsular Malaysia, Sumatra, Borneo, Palawan, Sulawesi and parts of the Moluccas.

SUBSPECIES The regional subspecies is *queda*.

HABITS AND HABITAT Fast-flying, and lands under leaves. Not uncommon in some localities. Inhabits forests in the lowlands.

TOP AND ABOVE: *Male*

Striated Angle ■ *Darpa striata* 1.7cm

DESCRIPTION Black-brown above, with tiny white forewing spots, paler veins and black streaks between the hindwing veins. Hindwing tornal area white, bearing 2 black marginal spots. Underside paler; hindwing more extensively white. The **Snowy Angle** *Darpa pteria* (highlands of Peninsular Malaysia and Thailand) lacks black spots on the white hindwing tornus. The **Hairy Angle** *Darpa hanria* (highlands of N Thailand) has wavy wing margins, a hairy and yellower hindwing tornus, and white forewing streaks instead of spots.

DISTRIBUTION NE India to Indo-China, Thailand, Peninsular Malaysia, Sumatra and Borneo.

SUBSPECIES Two subspecies: *striata* (Peninsular Malaysia, and S and SE Thailand) and *minta* (N Thailand).

HABITS AND HABITAT Fast-flying among low understorey vegetation. Rare in forests from low to moderate elevations.

Male of ssp. striata

Polygon Flat ■ *Odina hieroglyphica* 1.7cm

DESCRIPTION Unmistakable. Male deep orange with black borders and a network of black lines that give the appearance of blocks and streaks of orange. Female yellower.

Male of ssp. ortina

The **Zigzag Flat** *Odina decorata* (Continental Thailand) is orange on the hindwing and basal half of the forewing, with moderate-sized black spots. The outer half of the forewing is black with forked orange-grey streaks that give a zigzag appearance.

DISTRIBUTION N India to S Burma and S Thailand, Peninsular Malaysia, Singapore, Sumatra, Java and Borneo.

SUBSPECIES In Singapore and Peninsular Malaysia it is represented by subspecies *ortina* and in S Thailand by *ortygia*.

HABITS AND HABITAT Flies fast at bush height. Rare in lowland forests. More often seen in secondary forest, forest clearings and the forest edge.

Fulvous Pied Flat ■ *Pseudocoladenia dan* 1.5cm

DESCRIPTION Small. Reddish brown, chequered with indistinct darker brown spots. Forewing of male with 3 distinct and different-sized yellowish-white spots, the cell-end spot deeply cleft at its outer end. Tiny spots also present on the forewing costa and near the apex. Female with smaller, rounder spots, the cell-end spot divided or reduced to a single spot.

Male of ssp. dhyana

DISTRIBUTION NE India to S China, Thailand, Peninsular Malaysia, Singapore, Sumatra, Borneo and Java to Sulawesi.

SUBSPECIES Three subspecies with intergradations: *festa* (northernmost Thailand), *fabia* (the rest of N Thailand) and *dhyana* (south to Singapore).

HABITS AND HABITAT Fast-flying, preferring forest edges, clearings and riversides. Visits flowers of low herbaceous plants. Not uncommon in the right locations in lowland forest.

Tricolour Pied Flat ■ *Coladenia indrani* 1.8cm

DESCRIPTION Base of forewing and most of hindwing orange-brown above, spotted with black. Black outer two-thirds of forewing with prominent white discal and subapical spots as well as orange marginal spots. Hindwing black-bordered. Underside with orange areas more restricted, forming only streaks and spots. Readily recognised among 7 species of pied flats (*Coladenia*) in the region by its stark colouration. Most species are brown with large white forewing spots and small, diffuse black hindwing spots.

DISTRIBUTION India and Sri Lanka to Thailand (except the south) and Indo-China.
SUBSPECIES The ssp. *uposathra* occurs in this region.
HABITS AND HABITAT Fast-flying at low to moderate height, keeping to shady areas in the forest. Uncommon, occurring up to moderate elevations.

Male

Yellow Flat ■ *Mooreana trichoneura* 2.0cm

DESCRIPTION Sexes similar. A distinctive skipper with a yellow tornal area on a brown upperside. Forewing with small greyish-white spots and streaks. Both wings with grey-brown scaling and dark brown streaks between the veins, more prominently on the hindwing. Underside similar, but hindwing with the yellow area paler and more extensive, covering the lower two-thirds of the wing and turning white towards the base.

DISTRIBUTION N India to S China, Thailand, Peninsular Malaysia, Singapore, Sumatra, Java, Borneo and Palawan.
SUBSPECIES The two regional subspecies are *trichoneura* (Peninsular Malaysia and S Thailand) and *pralaya* (in the rest of Thailand).
HABITS AND HABITAT Conspicuous but fast in flight, keeping fairly low and landing under leaves. Uncommon in disturbed and primary lowland forests.

Male of ssp. trichoneura

SNOW FLATS – TAGIADES
Wings black-brown, usually with a white hindwing tornus above and more extensive white beneath. Forewing with tiny white spots. Both wings have rounder black spots than similar-looking *Darpa* (p.137). Eleven regional species. Fast-flying in bushes and low trees, landing on or under leaves.

Female of ssp. atticus

Common Snow Flat

■ *Tagiades japetus* 2.0cm

DESCRIPTION Lacks a white tornus in this region, but the hindwing underside may be overlaid with bluish-white scales.
DISTRIBUTION India to Indo-China, through Southeast Asia to Australia, New Guinea and the Solomon Islands.
SUBSPECIES Two subspecies: *atticus* (Singapore to S Thailand) and *ravi* (to the north) with a more bluish-white hindwing underside.
HABITS AND HABITAT Common in forested areas in the lowlands.

Large Snow Flat ■ *Tagiades gana* 2.1cm

DESCRIPTION Relatively large. Has 3 white dots near the forewing apex, and the white tornal patch does not surround the last black postdiscal spot above it. In the **Great Snow Flat** *Tagiades parra* (Thailand and Peninsular Malaysia), the white patch partially surrounds the black spot. The smaller **Burmese Snow Flat** *Tagiades ultra* (S Thailand to Singapore) has 7 white forewing dots, and a clear white hindwing tornus marked only with almost overlapping black marginal spots.
DISTRIBUTION India to Indo-China, Thailand, Peninsular Malaysia, Singapore, Sumatra, Java, Borneo and the Philippines.
SUBSPECIES Two subspecies: *gana* (Singapore to S Thailand) and *meetana* (further north), which barely has a white tornal patch in the wet season.
HABITS AND HABITAT Common in forest and the forest fringe in the lowlands.

Male of ssp. gana

Chestnut Angle ■ *Odontoptilum angulatum* 2.0cm

DESCRIPTION Wings angled at the termens. Brown above, dusted with white scales. Forewing with black-brown markings, a chestnut termen and small white spots. Hindwing crossed by whitish lines. The **Banded Angle** *Odontoptilum pygela* (lowlands of Peninsular Malaysia and S Thailand) is reddish brown above, whitened in the lower half of the

hindwing, and with white lines crossing both wings. The hindwing termen is strongly indented, giving the appearance of a triangular tail.

DISTRIBUTION Central India to S China, through Thailand, Peninsular Malaysia, Singapore, Sumatra, Java, Borneo and Palawan.

SUBSPECIES A single ssp. *angulatum* occurs in the region.

HABITS AND HABITAT Flies swiftly among low plants and small trees, usually landing on the upperside of leaves. Uncommon in lowland forest and at the forest edge.

Spotted Angle ■ *Caprona agama* 2.0cm

DESCRIPTION Brown or reddish brown above. Wet-season form with many yellowish-white spots on both wings. Hindwing white beneath, with black spots. Dry season form

very different: redder brown, with obscure dark spots, and with only several white spots on the forewing. The **Indian Skipper** *Spialia galba* (Continental Thailand) resembles the fully spotted form on the upperside but is much smaller, and the hindwing white markings are band-like.

DISTRIBUTION India to S China, including Continental Thailand. Absent from Peninsular Malaysia and Singapore but occurs in Java, the Lesser Sunda Islands and Sulawesi.

SUBSPECIES The subspecies in Thailand is *agama*.

HABITS AND HABITAT Flies moderately fast, keeping low and landing on ground-level vegetation. Common in forested areas.

Wet-season form

SWIFTS – HESPERIINAE
A large subfamily of skippers that land with wings closed, or with wings partially open when basking in sun. Unlike the preceding skipper subfamilies, their host plants are generally monocotyledons such as palms, bamboos and grasses.

Bush Hopper ■ *Ampittia dioscorides* 1.0cm

DESCRIPTION Small. Male brown above, with yellow markings comprising quadrangular forewing discal spots, a streak from the forewing base to mid costa, and a series of hindwing spots that form a partial band. Female with smaller and paler markings. Hindwing

yellowish beneath in both sexes, diffusely chequered with black-brown. Somewhat resembles a dart (p.158) but the forewing is spotted instead of banded and the spots are stacked instead of stepped.
DISTRIBUTION India to S China, through Thailand, Peninsular Malaysia, Singapore, Sumatra, Java, Borneo and Palawan.
SUBSPECIES Represented by ssp. *camertes* in this part of its range.
HABITS AND HABITAT Flies low. Easily overlooked by virtue of its small size and quick flight. Common, but local in occurrence, usually in forest clearings at low to moderate elevations.

Tiger Hopper ■ *Ochus subvittatus* 1.2cm

DESCRIPTION Black above, usually with a small yellowish-white spot near the forewing apex. Underside forewing apex and whole of hindwing orange-yellow with black streaks and spots; remainder of forewing black (usually covered by hindwing at rest).
DISTRIBUTION Burma and Thailand to S China. Does not extend into Peninsular Malaysia.
SUBSPECIES The single subspecies found in Thailand is *subvittatus*.
HABITS AND HABITAT Uncommon and local in occurrence in grassy areas near forest. Flies low, settling frequently.

A diverse genus, with 20 local species. Some are rare. Dark brown above with translucent white forewing spots. On the underside, the hindwing ranges from being unmarked to diffusely spotted to distinctly banded, and the forewing (and sometimes hindwing) bears characteristic diffuse, pale spots along the termen.

Dark Banded Ace
■ *Halpe ormenes* 1.6cm

DESCRIPTION Distinguished from other aces by its dark brown underside, overlaid with greenish-brown scales, and a prominent, creamy-white hindwing band.
DISTRIBUTION Burma, Thailand (where it is very scarce), Peninsular Malaysia, Singapore, Sumatra, Java, Borneo and Palawan.
SUBSPECIES The local subspecies is *vilasina*.
HABITS AND HABITAT Fast-flying, keeping to low plants and bushes. Uncommon in forest and at the forest edge in the lowlands.

Male

Light Straw Ace ■ *Pithauria stramineipennis* 2.1cm

DESCRIPTION Straw aces (*Pithauria*) are much larger than aces. The Light Straw Ace is black-brown above with yellowish-white forewing spots. Male with yellowish-green hair scales on forewing base and most of hindwing. Underside overlaid with buff brown scales, and with the reddish-brown under-scaling showing through as indistinct dark discal spots.

The **Branded Straw Ace** *Pithauria marsena* (lowlands of Thailand and Peninsular Malaysia) has yellowish-white spots on a browner underside.
DISTRIBUTION NE India to S China, Thailand, Peninsular Malaysia and Sumatra.
SUBSPECIES Represented by ssp. *stramineipennis*.
HABITS AND HABITAT Very swift-flying, but keeps low. Males settle on wet roads, rocks and riverbanks, and are fond of animal excreta. Localised, often occurring near rivers in forest up to moderate elevations.

Male

Forest Hopper ■ *Astictopterus jama* 1.5cm

DESCRIPTION Completely dark greyish brown, sometimes with tiny white costal spots near the forewing apex. Differentiated from the similar-looking Spotless Bob (p.145) by

its larger size, and from the Brown Bob (p.146) by a greyer brown colour and the absence of pale spots on the hindwing underside.

DISTRIBUTION NE India to S China, Thailand, Peninsular Malaysia, Singapore, Sumatra and Java.

SUBSPECIES Two subspecies occur in the region: *jama* (Singapore to southernmost Thailand) and *olivascens* (the rest of Thailand).

HABITS AND HABITAT Flies low at moderate pace, feeding on the flowers of herbaceous plants. Common in bushland near forest in the lowlands preferring more open habitats than the Spotless Bob and Brown Bob.

Ssp. jama

Chestnut Bob ■ *Iambrix salsala* 1.3cm

DESCRIPTION Small. Characterised by a rusty-brown underside that has small but distinct white discal spots and a white forewing cell spot. Upperside black-brown; forewing with a

row of indistinct pale discal spots in the male, which are white in the female. The smaller **Malayan Chestnut Bob** *Iambrix stellifer* (S and W Thailand to Singapore) has the white underside spots further from the hindwing cell, but is usually more easily distinguished by an additional white spot in the hindwing cell.

DISTRIBUTION India to S China, Thailand, Peninsular Malaysia, Singapore, Sumatra and Java.

SUBSPECIES The local subspecies is *salsala*.

HABITS AND HABITAT Flies moderately fast, usually among low herbaceous plants and bushes along forest roads and rivers. Common in forests in the lowlands, and sometimes even in villages and parks.

Small Red Bob ■ *Idmon obliquans* 1.3cm

DESCRIPTION The smallest orange-banded species of skipper in the region. Dark brown with a diffuse, orange-red postdiscal band parallel to the forewing termen. The band is short and does not cross the entire wing. The **Spotless Bob** *Idmon distanti* (lowlands of S Thailand and Peninsular Malaysia) is larger and lacks a red band – see also the Forest Hopper (p.144) and Brown Bob (p.146).

DISTRIBUTION S Burma, southernmost Thailand, Peninsular Malaysia, Sumatra, Java and Borneo.
SUBSPECIES The local subspecies is *obliquans*.
HABITS AND HABITAT Fairly fast-flying, frequenting herbaceous plants and bushes along forest paths and clearings. Common in secondary and primary forest in the lowlands.

Bright Red Velvet Bob ■ *Koruthaialos sindu* 1.7cm

DESCRIPTION Black brown with a bright orange-red forewing band that reaches the margin of the costa and almost reaches the tornus. Underside paler. Female with a wider and more orange band. In the slightly smaller **Narrow-banded Velvet Bob** *Koruthaialos rubecula* (Thailand and Peninsular Malaysia), the band is more restricted, not reaching the costa on the forewing upperside, and is absent in the race from SE Thailand. Unlike the Small Red Bob, the band is oblique.

DISTRIBUTION NE India to S China, through Thailand, Peninsular Malaysia, Sumatra, Java, Borneo and Palawan. Absent from Singapore.
SUBSPECIES Represented by ssp. *sindu* in this region.
HABITS AND HABITAT Moderately fast-flying, coming into clearings to feed at flowering herbaceous plants. Common in forested areas in the lowlands and highlands, especially near streams.

Brown Bob ■ *Psolos fuligo* 2.0cm

DESCRIPTION Wings brown on both sides, with indistinct pale spots beneath. Forewing with arched costa and often held low at rest, receding to the level of the hindwing and giving a hunched appearance. Forewing tip bent conspicuously outwards at rest. These and other characters differentiate it from other all brown skippers – see Forest Hopper (p.144) and Chocolate Demon.

Ssp. fuligo

DISTRIBUTION India to S China, through Thailand, Peninsular Malaysia, Sumatra, Java, Borneo, the Philippines and Sulawesi. Absent from Singapore.
SUBSPECIES The subspecies occurring in Peninsular Malaysia and S Thailand is *fuligo*, while ssp. *subfasciatus* is found further north in Thailand.
HABITS AND HABITAT Flies fairly fast at about bush height. Common in disturbed and primary forest in the lowlands, usually near streams and along forest roads.

Chocolate Demon
■ *Ancistroides nigrita* 2.3cm

Male of ssp. maura

DESCRIPTION Entirely brown, with the outer third of the undersides slightly paler. Female paler brown. Larger than the Brown Bob, lacks underside spots, and has a less hunched posture at rest.
DISTRIBUTION NE India to S China, through Thailand, Peninsular Malaysia, Singapore, Sumatra, Java, Borneo and the Philippines.
SUBSPECIES Two subspecies are found in this region: *maura*, in Singapore, Peninsular Malaysia and most of Thailand, and the smaller *diocles* in N Thailand.
HABITS AND HABITAT Fairly fast-flying, among low to moderately high bushes. Feeds from tubular flowers with its very long proboscis. Common in primary and secondary lowland forests, especially along rivers and forest roads wherever its host plant wild ginger grows.

Red Demon ▪ *Ancistroides armatus* 2.5cm

DESCRIPTION The largest of the orange-banded skippers. Readily distinguished from the Bright Red Velvet Bob (p.145) by its size. Dark brown, with an oblique, orange forewing band. Female with band paler and broader. The **Gem Demon** *Ancistroides gemmifer* (S Thailand and Peninsular Malaysia) is the size of the Bright Red Velvet Bob but the band is broader and less red, and the underside has indistinct spots that glisten purplish white at certain angles.
DISTRIBUTION S Burma, S Thailand, Peninsular Malaysia, Sumatra and Borneo.
SUBSPECIES Represented by ssp. *armatus*.
HABITS AND HABITAT Similar to the Chocolate Demon (p.146) but swifter in flight, preferring denser habitats. Inhabits forests up to the highlands.

Male

Common Banded Demon

▪ *Notocrypta paralysos* 2.0cm

DESCRIPTION One of 6 local species of banded demons (*Notocrypta*). All are black above, with a white band from costa to dorsum, and paler beneath with faint areas of whitish scaling. The Common Banded Demon has a conspicuously bent band. The **Spotted Demon** *Notocrypta feisthamelii* (N and S Thailand, and Peninsular Malaysia, including the highlands) has a series of tiny white spots near the forewing apex, and the white band reaches the underside costal margin.
DISTRIBUTION India to S China, through Southeast Asia as far east as the Philippines and Sulawesi.
SUBSPECIES Two local subspecies: *asawa* (NW Peninsular Malaysia and most of Thailand), and *varians* (S Thailand, the rest of Peninsular Malaysia, and Singapore).
HABITS AND HABITAT Similar to the Chocolate Demon (p.146) in behaviour and habitat. Visits flowers of low herbaceous plants.

ABOVE RIGHT AND RIGHT: *Ssp.* varians

Grass Demon ▪ *Udaspes folus* 2.3cm

DESCRIPTION An easily recognised skipper, with wings that look almost painted with water colours. Fairly large, with a rounded termen. Dark grey-brown above with several small to large white forewing spots and a large white hindwing patch. Wing margins thinly chequered black and white. Underside browner, especially along hindwing costa, and with additional patches of overlying white scales, especially between the white hindwing patch and wing base. Body white beneath.

DISTRIBUTION Sri Lanka and S India, Nepal to S China, Taiwan and Japan, through Thailand, Peninsular Malaysia, Singapore, Sumatra, Borneo, Java and the Lesser Sunda Islands.
SUBSPECIES Does not occur in any distinct subspecies across its range.
HABITS AND HABITAT Conspicuous, with a moderately fast and low flight. Lands with wings partly spread. Not uncommon in villages and parks and at the forest edge in the lowlands. Sometimes seen in gardens.

Sumatran Bob ▪ *Arnetta verones* 1.5cm

DESCRIPTION Small, with pointed forewings. Dark unmarked brown above. Dark brown below, with a slightly rusty look and a dull orange streak at the forewing apex, running parallel to the costa.

Mating pair

DISTRIBUTION S Thailand (where it is rare), Peninsular Malaysia, Sumatra and Borneo. Absent from Singapore.
SUBSPECIES None. It is uniform in appearance throughout its range.
HABITS AND HABITAT Flight varies from fast zipping when males guard territories, to gentle hopping when visiting flowers. Common where it occurs. Usually seen along forest clearings and the edges of forest roads, where it keeps to low herbaceous plants and bushes. Restricted to the lowlands.

Indian Palm Bob
■ *Suastus gremius* 1.7cm

DESCRIPTION Upperside dark brown with whitish forewing spots. Underside greyish brown with the forewing spots showing through and with 6 prominent black hindwing spots. Distinguished from the smaller Maculate Lancer (p.152) by a browner underside and fewer, larger black spots. Two smaller and rarer palm bobs (*Suastus*) that occur in the region have fainter black spots and whiter hindwing undersides.

DISTRIBUTION India to S China, Thailand, Peninsular Malaysia and Singapore. Also present in the islands of Sumba and Flores.
SUBSPECIES The subspecies occurring in this region is *gremius*.
HABITS AND HABITAT Fast-flying. It has become widespread through the cultivation of horticultural palms that are its host plant, and is now common in gardens and at the forest edge in the lowlands.

Female

Wax Dart ■ *Cupitha purreea* 1.5cm

DESCRIPTION Upperside dark brown with yellow costal and mid-wing streaks, as well as a yellow band across almost the width of the hindwing. Resembles a dartlet (p.158) above but appears completely yellow beneath at rest, although the forewing underside has a long black basal streak and a black tornus. Males have a hindwing cell pouch containing a waxy substance.

DISTRIBUTION S India, Nepal eastwards to Laos, and in Thailand, Peninsular Malaysia, Sumatra, Java, Borneo, Palawan and Sulawesi.
SUBSPECIES Occurs in a single form across its range without subspecies.
HABITS AND HABITAT Very fast-flying. Difficult to see until at rest. Usually flies low along forest paths and at the forest edge in the lowlands. Uncommon and localised in occurrence.

Purple and Gold Flitter ■ *Zographetus satwa* 1.6cm

DESCRIPTION One of 6 rare species of flitters of the genus *Zographetus* in the region. Yellow beneath, with a broad outer band of purple-brown, and with dark brown spots on

the hindwing yellow area. Above it is black-brown with small yellowish-white forewing spots. The rather different **Amber Flitter** *Zographetus doxus* (S Thailand to Singapore) has dark hindwing spots on a more uniform rich yellowish to orange-brown underside.

DISTRIBUTION NE India to S China, Thailand, Peninsular Malaysia and Java.

SUBSPECIES There are no subspecies as it is uniform across its range.

HABITS AND HABITAT Has a fast flight but lands often, keeping to low vegetation. Both species are rare in forested areas in the lowlands.

Male

Tree Flitter ■ *Hyarotis adrastus* 1.7cm

DESCRIPTION The 4 local species of flitters in the genus *Hyarotis* vary from being all black to having white markings. The Tree Flitter is the most strongly marked. The underside ground colour is black and brown. At rest with wings closed, it has the appearance of

an irregular white band from the forewing upper edge crossing the hindwing. The upperside is black-brown, with several medium- to small-sized white forewing spots.

DISTRIBUTION Sri Lanka and S India, N India to S China, through Thailand, Peninsular Malaysia, Singapore, Sumatra, Java, Borneo and the Philippines.

SUBSPECIES The subspecies found in this region is *praba*.

HABITS AND HABITAT Fast-flying among shrubs, bushes and low trees, usually near streams. Uncommon in forests and at the forest edge.

Male

Dubious Flitter ■ *Quedara monteithi* 1.8cm

DESCRIPTION Male almost black, and unmarked. Female also black, but with a white forewing band comprised of a squarish cell spot joined stepwise to a

Male

similar-sized spot beneath it, and to 2 small spots, including one just above the dorsum. In the rarer **Demon Flitter** *Oerane microthyrus* (S Thailand and Peninsular Malaysia),

both the male and the wider-banded female resemble the female Dubious Flitter, but are smaller, lack a spot above the dorsum, and have the antennal club whitened.
DISTRIBUTION NE India, Burma, S Thailand, Peninsular Malaysia, Singapore, Sumatra, Java, Borneo and the Philippines.
SUBSPECIES Occurs locally as ssp. *monteithi*.
HABITS AND HABITAT Very fast-flying and restless. Rare, inhabiting the understorey of lowland forest.

Female

Plain Tufted Lancer
■ *Isma iapis* 1.5cm

DESCRIPTION The genus *Isma* has 12 very similar-looking species in the region. They are grey-brown above with translucent white or yellowish forewing spots, and sometimes hindwing spots. Beneath, they are overlaid with paler scales. The Plain Tufted Lancer is yellowish brown beneath without white hindwing spots. The male has long hindwing tornal scales. The rather similar **Purple Tufted Lancer** *Isma protoclea* is darker beneath. The **Little Lancer** *Isma bononia* is the smallest, and has white hindwing spots. Both occur in S and SE Thailand, and Peninsular Malaysia.
DISTRIBUTION Burma, S Thailand, Peninsular Malaysia, Sumatra and Borneo.
SUBSPECIES The local subspecies is *iapis*.
HABITS AND HABITAT Fast-flying in the understorey and along forest paths and streams, usually settling only briefly. Occurs up to moderate elevations.

TOP AND ABOVE: *Male*

Silver-spotted Lancer ■ *Plastingia naga* 1.8cm

DESCRIPTION The underside is characteristic: black, with silvery-grey streaks and angular spots over the areas of the wings exposed at rest. Upperside dark brown with yellowish white forewing spots, and with dull orange-brown streaks on the forewing dorsum and hindwing. Body greyish white beneath, and abdomen striped black and white. In the rarer **Saffron-spotted Lancer** *Plastingia pellonia* (lowlands of S Thailand to Singapore),

the silvery-grey wing streaks and greyish body colours are instead rich orange-yellow, and the streaks on the upperside are more orange. DISTRIBUTION NE India to Laos, S and SE Thailand, Peninsular Malaysia, Singapore, Sumatra, Java, Borneo and the Philippines. SUBSPECIES No subspecies occur across its range. HABITS AND HABITAT Swift in flight, keeping to bushes and low trees in the understorey. Inhabits forest in the lowlands.

Male

Maculate Lancer ■ *Salanoemia sala* 1.8cm

DESCRIPTION Beneath the wings are greenish to purplish grey with at least 9 small dark spots on the hindwing and a few on the forewing apex. Above it is dark grey-brown with just a few translucent white forewing spots, and browner scaling on the hindwing. Four

other rarer species of *Salanoemia* occur in the region, all with dark hindwing spots. DISTRIBUTION SW India (Western Ghats), NE India, Burma, SE and southernmost Thailand, Peninsular Malaysia, Sumatra, Java, Borneo and Palawan. SUBSPECIES None. Occurs in a similar form throughout its range. HABITS AND HABITAT A fast flying butterfly. Not easily seen. Keeps fairly low along forest paths. Usually a rare inhabitant of lowland forest.

Male

Pugnacious Lancer ▪ *Pemara pugnans* 1.9cm

DESCRIPTION Rusty brown beneath and largely unmarked, except that the forewing spots show through faintly. Can be mistaken for a species of *Caltoris* (p.162), but on the less easily seen dark brown upperside it has larger and more elongate yellowish-white forewing spots, including a characteristic stacked double spot beyond the largest mid-wing spot. The upperside has some browner scaling on the forewing dorsum and on the hindwing.

DISTRIBUTION S Burma, S Thailand, Peninsular Malaysia, Singapore, Sumatra, Java and Borneo.

SUBSPECIES No subspecies occur.

HABITS AND HABITAT Swift in flight, generally taking to slightly higher vegetation in the understorey than *Caltoris* and related genera. Uncommon and localised in occurrence in both lowland and highland forest.

Male

Malayan Yellow-veined Lancer ▪ *Pyroneura latoia* 1.8cm

DESCRIPTION Underside striking: veins are highlighted yellow, and the spaces between are black, bearing bluish-white spots. Dark brown above with translucent white spots and yellow streaks on both wings. The 11 local species of lancers in the genus *Pyroneura* have the characteristic underside described, but the veins range from being yellow to red. The Malayan Yellow-veined Lancer can be differentiated from the rarer and rather similar **Sumatran Yellow-veined Lancer** *Pyroneura derna* (S Thailand and Peninsular Malaysia) by a bluish-white underside forewing stripe.

DISTRIBUTION S Burma, S and SE Thailand, Peninsular Malaysia, Singapore, Sumatra, Java and Borneo.

SUBSPECIES Occurs as ssp. *latoia*.

HABITS AND HABITAT Flies moderately fast in the understorey and along forest paths. Uncommon and local in occurrence. Inhabits forest up to moderate elevations.

Male

White-tipped Palmer
■ *Lotongus calathus* 2.0cm

DESCRIPTION Moderately large. Dark brown, with translucent white forewing spots above, which are separated in the male and larger and joined in the female. Underside with hindwing apex yellowish white, and with forewing spots forming a band that is extended to the costa and dorsum by yellowish scaling. At rest, it appears to have a straight band from forewing costa to hindwing apex.
DISTRIBUTION Burma to Indo-China, through Thailand, Peninsular Malaysia, Sumatra, Java, Borneo, Palawan and Sulawesi.

Male of ssp. balta (Cambodia)

SUBSPECIES Occurs as ssp. *calathus* from Peninsular Malaysia to S Thailand, and as ssp. *balta*, which has more developed spots, in the rest of Thailand.
HABITS AND HABITAT Very fast-flying. Keeps to the understorey and usually rests among thick vegetation. Rare in lowland forests.

Male of ssp. calathus

Giant Redeye ■ *Gangara thyrsis* 3.6cm

DESCRIPTION The largest skipper in the region. Brown above with 3 large yellowish forewing spots and 3 smaller spots nearer the apex. Underside with diffuse bands of bluish-grey scaling overlying the forewing apex and whole hindwing. In the smaller **Banded Redeye** *Gangara lebadea* (lowlands of Thailand to Singapore), the greyish scaling is more restricted and forms a band across the hindwing from apex to dorsum.
DISTRIBUTION India to S China, through Thailand, Peninsular Malaysia, Singapore, Sumatra, Java, Borneo, the Philippines and Sulawesi.
SUBSPECIES The regional subspecies is *thyrsis*.
HABITS AND HABITAT Fast-flying, but elusive in the day, being more active at twilight. Uncommon, inhabiting forests up to moderate elevations, and sometimes coming to the lights of buildings.

Male

Banana Skipper ■ *Erionota thrax* 3.4cm

DESCRIPTION Large. Brown above, with 3 yellowish quadrangular forewing spots, 2 large and 1 small. Paler beneath, dusted with buff-coloured scales on the hindwing and apical half of the forewing. In the very similar **Chinese Banana Skipper** *Erionota torus* (Thailand to Singapore), the forewing termen is more rounded. **Moore's Palm Redeye** *Erionota hiraca* (SE and S Thailand to Singapore) is smaller, with the upperside forewing apex whitened.
DISTRIBUTION NE India to Thailand, Peninsular Malaysia, Singapore and the Southeast Asian islands as far east as the Moluccas. Introduced to New Guinea and the Solomon Islands.
SUBSPECIES The local subspecies is *thrax*.
HABITS AND HABITAT Fast-flying. More active at twilight. Common in the lowlands wherever its host plants, wild or cultivated bananas grow. Considered a pest of bananas.

Male

Common Redeye ■ *Matapa aria* 1.8cm

DESCRIPTION Unmarked rusty orange-brown beneath. Hindwing tornus fringed with yellowish-grey scales. Brown above, without spots, but the male has a grey discal forewing stigma (a brand-like mark). Differentiated from swifts (p.159) by having red eyes. The **Dark-branded Redeye** *Matapa druna* (Thailand and Peninsular Malaysia) has the hindwing tornal fringe orange. A further 3 rarer species of redeyes occur in the region.
DISTRIBUTION Sri Lanka and India to S China, through Thailand, Peninsular Malaysia, Singapore, Sumatra, Java, the Lesser Sunda Islands, Borneo and the Philippines.
SUBSPECIES Without subspecies, as it is uniform across its range.
HABITS AND HABITAT Fast-flying and restless, but stays low. Uncommon, usually in open forests where bamboos grow.

Coconut Skipper ■ *Hidari irava* 2.8cm

DESCRIPTION Underside pale brown with a purplish tint. Unlike *Erionota* (Banana Skippers and Palm Redeyes, p.155), the forewing apical area and hindwing disc has small dark spots, some of which may be pale-centred, and there is usually a small, pale spot in the hindwing cell. Upperside brown with separated yellowish-white spots, the largest at the discs and cell-end, and the smallest near the apex. Unlike *Erionota*, there is a spot above the dorsum.

Male

DISTRIBUTION India to S and SE Thailand, Peninsular Malaysia, Singapore, Sumatra, Java, Borneo and southern Philippines.
SUBSPECIES Occurs in a single form without subspeciation.
HABITS AND HABITAT Fast-flying. More active at twilight. Common, and found in a variety of lowland habitats, especially where coconut, one of its host plants, is grown.

Whitespot Palmer ■ *Eetion elia* 2.5cm

DESCRIPTION Forewing elongate. Black-brown above. Forewing with a band of well-separated white spots running from the dorsum to near the apex, and with some small costal spots. Hindwing with a discal row of narrow quadrangular spots. Underside brown, slightly chocolate on the hindwing, with the upperside spots visible and with a prominent white triangular band across the hindwing, leaving the costa and termen brown.

Male

DISTRIBUTION Burma, S Thailand, Peninsular Malaysia, Singapore, Sumatra and Borneo.
SUBSPECIES None. It does not vary much with geographical location.
HABITS AND HABITAT Very fast-flying and restless, often settling in dense foliage. Generally uncommon, but several may be seen where it occurs. Inhabits the forest understorey in the lowlands.

Striped Green Palmer ■ *Pirdana hyela* 2.0cm

DESCRIPTION Underside black with metallic bluish-green stripes along the veins. Body and head orange beneath. Hindwing with an orange tornus, lacking black markings (unlike awl kings, p.136). Upperside black-brown, with the bases shining greenish blue (more extensive in the female). The **Plain Green Palmer** *Pirdana distanti* (lowlands of SE and S Thailand to Peninsular Malaysia) has a more uniformly dull metallic green underside.

DISTRIBUTION NE India to Indo-China, Thailand, Peninsular Malaysia, Sumatra, Java, Borneo, Palawan, N Philippines (Luzon) and Sulawesi.
SUBSPECIES The local subspecies is *rudolphii*.
HABITS AND HABITAT Swift in flight, keeping to low vegetation and landing on exposed leaves. Usually rare, inhabiting forest near streams and swamps at low to moderate elevations.

Yellow Grass Dart
■ *Taractrocera archias* 1.1cm

DESCRIPTION One of 5 local grass darts (*Taractrocera*). They are orange to yellow, and black. Unlike other skippers, they lack a pointed hook on the clubbed antennal tip. The Yellow Grass Dart has overlapping orange-yellow forewing spots tapering in width towards the apex. The orange costal streak usually touches the mid-wing spots. The **Malayan Grass Dart** *Taractrocera ardonia* (lowlands of Peninsular Malaysia and Singapore) is distinctive with its small creamy-yellow spots and yellowish-brown underside.

DISTRIBUTION S Burma, Thailand, Indo China, Peninsular Malaysia, Singapore, Java and the Lesser Sunda Islands.
SUBSPECIES The local subspecies is *quinta*.
HABITS AND HABITAT Flies fast and low among grasses and shrubs. Common but localised, occurring in villages, open country and the forest edge in the lowlands.

TOP AND ABOVE: *Male*

Common Dartlet ■ *Oriens gola* 1.3cm

DESCRIPTION Black above, with orange bands. Orange-brown beneath, with pale bands diffusely edged black. Dartlets (*Oriens*) generally have the forewing band broader at midpoint than in darts and palm darts (p.159), touching the cell markings above it. The **Indian Dartlet** *Oriens goloides* (lowlands of Thailand and Peninsular Malaysia) has the veins darkened across the upperside bands. The **Malayan Dartlet** *Oriens paragola* (lowlands of S Thailand and Peninsular Malaysia) is brown beneath and lacks black edging to the hindwing band.
DISTRIBUTION NE India to S China, Central, SE and S Thailand to Singapore, Java and the Lesser Sunda Islands, and Sumatra eastward to North Philippines.
SUBSPECIES The local subspecies is *pseudolus*.
HABITS AND HABITAT Fast-flying. Occurs in secondary forest and at the forest edge up to moderate elevations.

DARTS – POTANTHUS
Similar to dartlets but the forewing band is separated from the cell markings. Nineteen similar-looking species occur in the region.

Male

Lesser Dart ■ *Potanthus omaha* 1.2cm

DESCRIPTION Characterised by darkened veins crossing the orange markings. The **Lesser Orange Dart** *Potanthus ganda* is deeper orange, without darkened veins. The larger **Broad Bident Dart** *Potanthus trachala* is yellower, with the three lowermost forewing discal spots concave and detached from the spots beyond them. Both occur in Thailand and Peninsular Malaysia.
DISTRIBUTION Central Burma to Vietnam, through Thailand, Peninsular Malaysia, Singapore, Sumatra, Borneo, the Philippines and Sulawesi.
SUBSPECIES Represented by ssp. *omaha*.
HABITS AND HABITAT Flies fast among shrubs, grasses and bushes. Common at the verges of lowland forest.

Male

PALM DARTS – *TELICOTA*
Larger than darts and dartlets (p.158). Males have a greyish forewing brand. The 7 local species may be difficult to differentiate.

Pale Palm Dart ■ *Telicota colon* 1.4cm

DESCRIPTION Male black above, with orange-yellow bands and characteristic yellow-streaked veins along the forewing termen. Underside largely orange-yellow. Female yellower with forewing band scalloped. The **Chinese Palm Dart** *Telicota besta* (Thailand to Singapore) is similar in colour but lacks streaks and scallops. The **Bright Orange Palm Dart** *Telicota augias* (Central Thailand to Singapore) is more orange beneath than other species.

DISTRIBUTION India to S China, through Southeast Asia (including Singapore), Australia, New Guinea and the Solomon Islands.

SUBSPECIES The local subspecies is *stinga*.

HABITS AND HABITAT Fast-flying among tall grasses and shrubs. Common at the forest edge, waysides and even gardens in the lowlands.

ABOVE RIGHT AND RIGHT: *Male*

SWIFTS – BAORINI
A tribe of mostly brown skippers. Species often difficult to differentiate. In this region, they comprise 35 species in the genera *Parnara*, *Borbo*, *Pelopidas*, *Polytremis*, *Baoris*, *Caltoris* and *Iton*.

Ceylon Swift ■ *Parnara bada* 1.4cm

DESCRIPTION Small. Antennae short. Brown above, and grey-brown beneath with yellowish-brown scaling. Forewing with 2 white discal spots, and a few dots near the apex, but with none above the dorsum or in the cell. Hindwing with a row of 4 or more small white spots. All spots visible beneath.

DISTRIBUTION India to S China, through Southeast Asia (absent in Singapore), to Timor and Eastern Australia.

SUBSPECIES Represented by ssp. *bada*.

HABITS AND HABITAT Swift, keeping low. Locally common in coastal areas, waysides and forest edges up to the highlands.

Rice Swift
■ *Borbo cinnara* 1.5cm

DESCRIPTION Sometimes called the Formosan Swift. Brown beneath and above, dusted with greenish- to yellowish-brown scales. Forewing with white spots, including a spot above the dorsum and sometimes small cell spots. Hindwing with 3 small white rounded spots that are more distinct beneath.
DISTRIBUTION India to S China, throughout Southeast Asia and as far as Australia, New Guinea, the Solomon Islands and the New Hebrides.
SUBSPECIES No subspecies occur.
HABITS AND HABITAT Fast-flying, usually around grasses, shrubs and bushes. Common in secondary growth and open country up to moderate elevations. Rice is one of its host plants.

LEFT: *Male*

Small Branded Swift ■ *Pelopidas mathias* 1.5cm

DESCRIPTION Brown above with a slightly golden to green sheen near the wing bases. Male with white forewing spots, including 2 cell spots, and with a brand that touches the outer edge of the lowest spot. Female with larger spots, and with an additional spot above the dorsum and often one just below the largest spot, but rarely with hindwing spots. Underside grey-brown, with at least 4 white hindwing spots. The very similar **Bengal Swift** *Pelopidas agna* (Thailand to Singapore) is browner beneath and the male brand touches the white spot centrally.
DISTRIBUTION Africa, Turkey and Arabia, through India eastward to Japan, throughout Southeast Asia and New Guinea.
SUBSPECIES The subspecies *mathias* occurs in this region.
HABITS AND HABITAT Swift, keeping to low bushes, shrubs and grasses. Common in open country and secondary forest up to the highlands.

Male

Conjoined Swift ■ *Pelopidas conjunctus* 2.1cm

DESCRIPTION Larger and more rusty brown than the preceding swifts. Upperside slightly reddish brown with moderately large yellowish forewing spots, including equal-sized cell spots and a spot above the dorsum. Hindwing with a few small spots. Underside paler, with hindwing spotted. The even larger but rarer **Great Swift** *Pelopidas assamensis* (Thailand to Singapore) is blacker

with white spots, and the cell spots are partly joined.

Female

DISTRIBUTION India to S China, through Thailand, Peninsular Malaysia, Singapore, Sumatra, Borneo, the Philippines, Java and the Lesser Sunda Islands.
SUBSPECIES Locally represented by ssp. *conjunctus*.
HABITS AND HABITAT A fast flier, coming to shrubs and bushes at the edges of forests and along forest roads. Fairly common up to the highlands.

Contiguous Swift
■ *Polytremis lubricans* 1.7cm

DESCRIPTION Dark golden brown above with translucent yellowish-white forewing spots, including a small spot above the dorsum and 2 cell spots that may be joined. Hindwing with 2 tiny yellowish spots. Underside rusty orange-brown; hindwing with small and sometimes indistinct spots. The **Yellow-spotted Swift** *Polytremis eltola* (Thailand and Peninsular Malaysia, in the highlands) is larger, with rounder, yellower spots and prominent underside hindwing spots.
DISTRIBUTION India to S China, through Thailand, Peninsular Malaysia, Singapore, Sumatra, Java, Borneo, the Sulu Islands (Sibutu and Tawi Tawi), Sulawesi and Timor.
SUBSPECIES The subspecies occurring in this region is *lubricans*.
HABITS AND HABITAT Quick-flying, resting on shrubs and low vegetation. Commonly inhabits open areas in forests, as well as villages and parks in the lowlands.

TOP AND ABOVE: *Male*

TOP AND ABOVE: *Male*

Fullstop Swift ■ *Caltoris cormasa*
1.8 cm

DESCRIPTION The 13 local species of *Caltoris* are medium-sized and dark brown above, with whitish forewing spots. Hindwing spots are rare and, if present, only occur on the underside. The Fullstop Swift is rusty brown beneath and usually has a single forewing cell spot. The larger **Colon Swift** *Caltoris bromus* is more yellow beneath, while the **Lesser Colon Swift** *Caltoris cahira* is dark chocolate brown beneath. Both have 2 unequally sized forewing cell spots, and occur in Thailand and Peninsular Malaysia.
DISTRIBUTION N India, Burma, S Thailand, Peninsular Malaysia, Singapore, Sumatra, Java, Borneo and the Philippines.
SUBSPECIES No subspecies occur.
HABITS AND HABITAT Quick in flight, settling inconspicuously on low vegetation at the forest edge. Moderately common in lowland forest.

Moore's Wight ■ *Iton semamora* 2.1cm

DESCRIPTION Fairly large. Very distinctive on the underside when at rest: chocolate brown on the forewing and costa of the hindwing, with a sharp, straight-line transition to white on the rest of the hindwing and lower body. Underside termens with black

angular spots, which are prominent on the white area. Upperside dark brown with small whitish forewing spots, and with the hindwing tornus greatly whitened.
DISTRIBUTION NE India to Indo-China, through Thailand, Peninsular Malaysia, Sumatra and Borneo.
SUBSPECIES The local subspecies is *semamora*.
HABITS AND HABITAT Flies fairly fast. The white makes it very conspicuous in flight. Keeps to low vegetation in or near forests. Usually rare. Occurs at low to moderate elevations.

Female

CLASSIFICATION

The following list is a classification of butterflies found in Thailand, Peninsular Malaysia and Singapore, from their superfamilies down to the level of genera. It is intended to give an understanding of the relationships of butterflies, and may not always follow the arrangement of the species descriptions in this book (see pp.17–18 for further explanation). The classification is based primarily on *Butterflies of the Malay Peninsula* (1992) and its update article, as well as *Butterflies in Thailand* Vol.1–6 (1981–1996), and the *Tree of Life Web Project* (www.tolweb.org) together with the various articles it cites.

Taxonomic classifications are based on theories of the evolutionary lineages of species, and may be derived from taxonomists' intuition or an analysis of the characteristics of species. Analyses of morphological characters and gene sequences produce branching tree-like diagrams called phylogenetic trees, which are used to interpret the relationships of species and their descents from common ancestors. Such trees give us an understanding of the relatedness of species and higher taxa (i.e., higher groupings). Species and higher taxa that share the same branches share greater similarity. The classifications produced from the trees are hierarchical, that is, taxa are nested or grouped in higher taxa, which are nested in yet higher taxa. The highest taxa are usually determined by the lower branching points in the diagrams, with other higher taxa forming branches further up the tree. Each higher taxon is split into one or more taxa under it that normally share the same rank as each other. For example, the subfamily Lycaeninae is split into two tribes, Lycaenini and Heliophorini, and there are no genera outside these tribes. In recent years, however, some taxonomists have moved away from equal rank classifications because the branching patterns of evolutionary lineages suggest outlying genera. An example is *Rhinopalpa* in the subfamily Nymphalinae, which is currently not assigned to any of the three tribes in Nymphalinae.

The higher taxa above the rank of genera that are governed by the International Code of Zoological Nomenclature and its Commission are superfamilies, families, subfamilies, tribes and subtribes. However, other groupings are also used by taxonomists. For example, where the exact relationships of genera are uncertain or where there is a need to subdivide higher taxa further, groups (or sections) may be used. These are usually named after the oldest named species or genus in the group, for example the *Nacaduba* group, which in this region comprises the genera *Nacaduba*, *Ionolyce*, *Prosotas* and *Catopyrops*. Butterflies themselves belong to an informal grouping that is referred to as Rhopalocera.

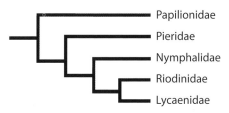

Phylogenetic tree for families in the superfamily Papilionoidea.

When a classification is shown in the form of a list, as below, some of the information on relationships cannot be shown. The branches with the least sub-branches on the phylogenetic tree, or in other words the most basal branches, are usually listed first. However, the order of species, genera and higher taxa in a listing is to some degree subjective, especially when there are equal branches on the tree. For this reason, classificatory listings may differ to some extent in their order.

Classification is a continuously ongoing work, as the amount of information needed to produce a good classification is enormous. It is therefore common for classifications to be revised frequently based on a more comprehensive morphological analysis or a new gene analysis or a consensus of different analyses. In some groups of butterflies that have been less well studied, the classification is still very tentative, and genera have not yet been assigned to formal taxon groups. This is especially noticeable in the Hesperiinae and in some taxa of Lycaenidae. The following classification is, therefore, an update of a work in progress, but it will not remain static. With newer studies in the future, the classification is expected to change to some degree. These changes, however, will continue to shed new light on the relationships of butterflies.

SUPERFAMILY HESPERIOIDEA

Now considered to comprise a single family, Hesperiidae.

FAMILY HESPERIIDAE

Subfamily Coeliadinae
Badamia
Bibasis
Burara
Hasora
Choaspes

Subfamily Eudaminae
Lobocla

Subfamily Pyrginae
Tribe Celaenorrhinini
Celaenorrhinus
Pseudocoladenia
Sarangesa
Tribe Tagiadini
Capila
Tapena
Darpa
Odina
Coladenia
Satarupa
Seseria
Pintara
Chamunda
Gerosis
Abraximorpha
Mooreana
Tagiades
Ctenoptilum
Odontoptilum
Caprona
Tribe Carcharodini
Spialia

Subfamily Hesperiinae
A few tribes are recognised, but the majority of genera in this subfamily have only been assigned to loose, informal groups of genera that may not correctly reflect their relationships, especially in the *Plastingia* group of genera.
Tribe Aeromachini
Aeromachus
Ampittia
Ochus
Sebastonyma
Sovia

Onryza
Thoressa
Halpe
Pithauria
Ancistroides group
    Astictopterus
    Iambrix
    Idmon
    Koruthaialos
    Psolos
    Stimula
    Ancistroides
    Notocrypta
    Udaspes
Plastingia group
    Arnetta
    Suada
    Scobura
    Suastus
    Cupitha
    Zographetus
    Oerane
    Hyarotis
    Quedara
    Isma
    Xanthoneura
    Plastingia
    Salanoemia
    Pemara
    Pyroneura
    Pseudokerana

Lotongus
Zela
    Subgenus *Zela*
    Subgenus *Matapoides*
    Subgenus *Zampa*
Gangara
Erionota
Ge
Matapa
Unkana
Hidari
Eetion
Acerbas
Pirdana
Creteus
Tribe Baorini
    Parnara
    Borbo
    Pelopidas
    Polytremis
    Baoris
    Caltoris
    Iton
Tribe Taractrocerini
    Taractrocera
    Oriens
    Potanthus
    Cephrenes
    Telicota
Tribe Hesperiini
    Ochlodes

SUPERFAMILY PAPILIONOIDEA

FAMILY PAPILIONIDAE

Subfamily Parnassiinae

Bhutanitis lidderdalii, the sole representative of this subfamily in the region, was once found in Thailand as an endemic subspecies, *ocellatomaculata*, which is now thought to be extinct.

    Tribe Zerynthiini
        Bhutanitis

Subfamily Papilioninae

Species once placed in *Paranticopsis* are now included in the subgenus *Pathysa*.

    Tribe Leptocircini
        Lamproptera
        Graphium
            Subgenus *Pazala*
            Subgenus *Pathysa*
            Subgenus *Graphium*

SUPERFAMILY PAPILIONOIDEA cont.

Tribe Teinopalpini
Teinopalpus
Meandrusa
Tribe Troidini
Trogonoptera
Troides
Atrophaneura
Byasa
Losaria
Pachliopta
Tribe Papilionini
Papilio
Subgenus *Princeps*
Subgenus *Chilasa*

FAMILY PIERIDAE

Subfamily Coliadinae
Eurema
Subgenus *Terias*
Subgenus *Eurema*
Gandaca
Dercas
Catopsilia
Colias

Subfamily Pierinae
A number of genera in the subfamily
are unassigned to tribes. *Hebomoia*,
Pareronia and *Ixias* belong to a division
referred to as the *Colotis* group.
Hebomoia
Pareronia
Ixias
Leptosia
Tribe Pierini
Subtribe Appiadina
Saletara
Appias
Phrissura
Subtribe Pierina
Talbotia
Pontia
Pieris
Subgenus *Artogeia*

Subtribe Aporiina
Cepora
Prioneris
Aporia
Delias

FAMILY NYMPHALIDAE

Subfamily Libytheinae
Libythea

Subfamily Danainae
Tribe Danaini
Subtribe Danaina
Danaus
Subgenus *Anosia*
Subgenus *Salatura*
Tirumala
Parantica
Ideopsis
Subgenus *Radena*
Subgenus *Ideopsis*
Subtribe Euploeina
Idea
Euploea

Subfamily Calinaginae
Calinaga

Subfamily Satyrinae
Coelites is tentatively placed in the
subtribe Eritina but recent studies
suggest it is more closely related to
the subtribe Eutychiina, which mainly
comprises species from the Americas.
Tribe Elymniini
Elymnias
Tribe Zetherini
Xanthotaenia
Neorina
Penthema
Ethope
Tribe Amathusiini
Faunis
Melanocyma
Taenaris

Aemona
Stichophthalma
Amathusia
Amathuxidia
Zeuxidia
Thaumantis
Thauria
Discophora
Enispe
Tribe Satyrini
Subtribe Eritina
Coelites
Erites
Orsotriaena
Subtribe Ragadiina
Ragadia
Subtribe Parargina
Orinoma
Subtribe Mycalesina
Mycalesis
Subtribe Lethina
Lethe
Neope
Mandarinia
Subtribe Ypthimina
Callerebia
Ypthima
Tribe Melanitini
Melanitis

Subfamily Charaxinae
Tribe Charaxini
Polyura
Charaxes
Tribe Prothoini
Prothoe
Agatasa

Subfamily Heliconiinae
Species sometimes placed under
Childrena and *Argyreus* are included in
Argynnis.
Tribe Cethosiini
Cethosia
Tribe Heliconiini
Dryas

Tribe Acraeini
Acraea
Tribe Argynnini
Subtribe Argynnina
Argynnis
Tribe Vagrantini
Vindula
Algia
Cirrochroa
Terinos
Phalanta
Vagrans
Cupha

Subfamily Limenitidinae
Athyma is sometimes split into *Athyma*
and *Tacola*. *Tanaecia* includes species
sometimes separated under *Cynitia*.
Tribe Parthenini
Parthenos
Lebadea
Bhagadatta
Tribe Limenitidini
Sumalia
Parasarpa
Auzakia
Moduza
Pandita
Athyma
Tribe Neptini
Pantoporia
Lasippa
Neptis
Phaedyma
Tribe Adoliadini
Subtribe Adoliadina
Bassarona
Dophla
Euthalia
Tanaecia
Lexias
Neurosigma

Subfamily Pseudergolinae
Dichorragia
Stibochiona
Pseudergolis

SUPERFAMILY PAPILIONOIDEA cont.

Subfamily Apaturinae
Species placed under *Mimathyma* and
Chitoria are sometimes lumped under
the genus *Apatura*.

> Mimathyma
> Dilipa
> Chitoria
> Rohana
> Helcyra
> Eulaceura
> Herona
> Euripus
> Sephisa
> Hestina

Subfamily Biblidinae
Tribe Biblidini
> Ariadne
> Laringa

Subfamily Cyrestinae
> Cyrestis
> Chersonesia

Subfamily Nymphalinae
Rhinopalpa is related to African
Vanessula. Both are currently
unassigned to a tribe.
Tribe Nymphalini
> Vanessa
> Symbrenthia
> Polygonia
> Kaniska

> Rhinopalpa

Tribe Junoniini
> Hypolimnas
> Junonia
> Yoma
Tribe Kallimini
> Doleschallia
> Kallima

FAMILY RIODINIDAE

Subfamily Nemeobiinae
> Zemeros
> Dodona
> Stiboges
> Abisara
> Paralaxita
> Laxita
> Taxila

FAMILY LYCAENIDAE
In the large tribe Polyommatini and large
subfamily Theclinae, informal groups of
genera are currently used. These may
eventually be re-sorted into formal
groupings such as tribes or subtribes.

Subfamily Curetinae
> Curetis

Subfamily Poritiinae
Tribe Poritiini
> Cyaniriodes
> Poritia
> Simiskina
> Deramas

Subfamily Miletinae
Tribe Liphyrini
> Liphyra
Tribe Miletini
> Miletus
> Allotinus
>> Subgenus *Allotinus*
>> Subgenus *Fabitaras*
>> Subgenus *Parayerydus*
> Logania
Tribe Spalgini
> Taraka
> Spalgis

Subfamily Aphnaeinae
> Spindasis

Subfamily Polyommatinae

Everes in the *Cupido* group is sometimes merged into the genus *Cupido*.

Tribe Polyommatini
- *Castalius* group
  - *Castalius*
  - *Tarucus*
- *Upolampes* group
  - *Discolampa*
  - *Caleta*
- *Cupido* group
  - *Everes*
  - *Talicada*
  - *Tongeia*
  - *Bothrinia*
- *Pithecops* group
  - *Pithecops*
- *Azanus* group
  - *Azanus*
- *Lycaenopsis* group
  - *Lycaenopsis*
  - *Neopithecops*
  - *Megisba*
  - *Oreolyce*
  - *Cebrella*
    - Subgenus *Cebrella*
    - Subgenus *Chelakina*
  - *Lestranicus*
  - *Plautella*
  - *Callenya*
  - *Acytolepis*
  - *Udara*
    - Subgenus *Udara*
    - Subgenus *Selmanix*
    - Subgenus *Penudara*
  - *Celastrina*
  - *Celatoxia*
  - *Monodontides*
- *Glaucopsyche* group
  - *Caerulea*
- *Zizeeria* group
  - *Zizina*
  - *Zizeeria*
- *Zizula* group
  - *Zizula*
- *Famegana* group
  - *Famegana*
- *Polyommatus* group
  - *Chilades*
- *Euchrysops* group
  - *Euchrysops*
- *Catochrysops* group
  - *Catochrysops*
- *Lampides* group
  - *Lampides*
- *Leptotes* group
  - *Leptotes*
- *Jamides* group
  - *Jamides*
- *Nacaduba* group
  - *Nacaduba*
  - *Ionolyce*
  - *Prosotas*
  - *Catopyrops*
- *Petrelaea* group
  - *Petrelaea*
- *Una* group
  - *Una*
  - *Orthomiella*

Tribe Niphandini
- *Niphanda*

Tribe Lycaenesthini
- *Anthene*

Subfamily Lycaeninae

Tribe Heliophorini
- *Heliophorus*

Subfamily Theclinae

Chrysozephyrus includes species sometimes placed under *Shirozuozephyrus*. *Virachola* is sometimes considered to be a subgenus of *Deudorix*. *Sinthusa* includes species sometimes placed under *Pseudochliaria*.

Tribe Luciini
- *Hypochrysops* group
  - *Hypochrysops*

Tribe Theclini
- *Thecla* group
  - *Ussuriana*

SUPERFAMILY PAPILIONOIDEA cont.

Austrozephyrus
Chrysozephyrus
Tribe Arhopalini
  Arhopala group
    Arhopala
    Flos
    Thaduka
    Mahathala
    Apporasa
    Neomyrina
  Semanga group
    Semanga
    Mota
  Surendra group
    Surendra
      Subgenus *Surendra*
      Subgenus *Zinaspa*
Tribe Amblypodiini
  Iraota
  Amblypodia
Tribe Catapaecilmatini
  Catapaecilma
  Acupicta
Tribe Loxurini
  Drina group
    Drina
  Loxura group
    Loxura
    Yasoda
    Eooxylides
    Thamala
Tribe Cheritrini
  Cheritra group
    Cheritra
    Ticherra
    Cheritrella
    Ritra
    Cowania
    Drupadia
Tribe Horagini
  Horaga
Tribe Iolaini
  Britomartis group
    Britomartis

Iolaus group
  Dacalana
  Pratapa
  Creon
  Tajuria
  Bullis
  Purlisa
  Rachana
    Subgenus *Rachana*
  Jacoona
  Neocheritra
  Thrix
  Mantoides
  Manto
  Charana
  Suasa
Tribe Remelanini
  Remelana
  Pseudotajuria
  Ancema
Tribe Hypolycaenini
  Hypolycaena
  Zeltus
  Chliaria
Tribe Eumaeini
  Satyrium group
    Satyrium
  Callophrys group
    Callophrys
Tribe Deudorigini
  Capys group
    Pamela
  Deudorix group
    Deudorix
    Virachola
    Artipe
    Sinthusa
    Bindahara
    Rapala
    Araotes
    Sithon

Further Reading

Corbet, A.S. and Pendlebury, H.M. (1992). *The Butterflies of the Malay Peninsula*. 4th edn. Revised by J.N. Eliot. Malayan Nature Society.

D'Abrera, B. (1982). *Butterflies of the Oriental Region. Part I. Papilionidae, Pieridae & Danaidae*. Hill House (in association with E.W. Classey).

D'Abrera, B. (1984). *Butterflies of the Oriental Region. Part II. Nymphalidae, Satyridae & Amathusidae*. Hill House (reprinted 1985).

D'Abrera, B. (1986). *Butterflies of the Oriental Region. Part III. Lycaenidae and Riodinidae*. Hill House.

Davison, G.W.H., WWF Malaysia and Cubitt, G.S. (2003). *The National Parks and Other Wild Places of Malaysia*. New Holland.

Ek-Amnuay, P. (2012). *Butterflies of Thailand*. 2nd edn. Amarin Printing and Publishing.

Eliot, J.N. (2006). Updating *The Butterflies of the Malay Peninsula*. Edited, enlarged and prepared for publication by H. Barlow, R. Eliot, L.G. Kirton and R.I. Vane-Wright. *Malayan Nature Journal* 59: 1–49 (also available as bound separate offprint).

Fleming, W.A. (1983). *Butterflies of West Malaysia and Singapore*. 2nd edn. Revised by A. McCartney. Longman.

Gan, C.W. and Chan, S.K.M. (2007). *A Field Guide to the Butterflies of Singapore*. Butterfly Interest Group, Nature Society (Singapore).

Igarashi, S. and Fukuda, H. (1997). *The Life Histories of Asian Butterflies*. Volume 1. Tokai University Press.

Igarashi, S. and Fukuda, H. (2000). *The Life Histories of Asian Butterflies*. Volume 2. Tokai University Press.

Khew, S.K. (2010). *A Field Guide to the Butterflies of Singapore*. Ink on Paper Communications.

Kimura, Y., Aoki, T., Yamaguchi, S., Uémura, Y. and Saito, T. (2011). *The Butterflies of Thailand. Based on Yunosuke Kimura Collection*. Volume 1. Hesperiidae · Papilionidae · Pieridae. Mokuyosha.

Lekagul, B., Askins, K., Nabhitabhata, J. and Samruadkit, A. (1977). *Field Guide to the Butterflies of Thailand*. Association for the Conservation of Wildlife.

Maxwell, J.F. (2013). A reassessment of the forest types of Thailand. Pp. 1–17 in: *Vegetation and Vascular Flora of Doi Sutep-Pui National Park,*

Northern Thailand (eds J.F. Maxwell and S. Elliot). Thai Studies in Biodiversity No. 5. Biodiversity Research and Training Program.

Morrell, R. (1960). *Common Malayan Butterflies*. Illustrated by A.H. Burvill. Longman (reprinted 1982).

National Park Wildlife and Plant Conservation Department (Thailand). (2004). *The Best of National Parks of Thailand*. National Park Office, National Park, Wildlife and Plant Conservation Department (Thailand) (www.dnp.go.th/parkreserve/e-book/book/np_best_en. pdf?).

Neo, S.S.H. (1996). *A Guide to Common Butterflies of Singapore*. Singapore Science Centre.

Otsuka, K. (2001). *A Field Guide to the Butterflies of Borneo and South East Asia*. Hornbill Books.

Pinratana, A. (1981). *Butterflies in Thailand. Volume 4. Lycaenidae*. Viratham Press.

Pinratana, A. (1983). *Butterflies in Thailand. Volume 2. Pieridae and Amathusiidae*. Viratham Press.

Pinratana, A. (1985). *Butterflies in Thailand. Volume 5. Hesperiidae*. Viratham Press.

Pinratana, A. (1988). *Butterflies in Thailand. Volume 6. Pieridae and Amathusiidae*. Viratham Press.

Pinratana, A. (1996). *Butterflies in Thailand. Volume 3. Nymphalidae*. 2nd edn. Revised. Brothers of St. Gabriel in Thailand.

Pinratana, A. and Eliot, J.N. (1992). *Butterflies in Thailand. Volume 1. Papilionidae and Danaidae*. Brothers of St. Gabriel in Thailand.

Tan, H. and Khew, S.K. (2012). *Caterpillars of Singapore's Butterflies*. (With an unfolding pocket insert Identification Guide.) National Parks Board (Singapore).

Whitmore, T.C. (1984). *Tropical Rain Forests of the Far East*. 2nd edn. Clarendon Press.

Yong, H.-S. (1983). *Malaysian Butterflies – An Introduction*. Tropical Press.

Web Resources

A Checklist of Butterflies in Indo-China
http://yutaka.it-n.jp
Yutaka Inayoshi

Butterflies of Singapore. Featuring Nature's Flying Jewels in Singapore!
www.butterflycircle.blogspot.com
Butterfly Circle

▪ Further Information ▪

Butterfly Interest Group (BIG)
www.butterfly.nss.org.sg
Nature Society (Singapore)

Chin's Nature Corner
www.oocities.org/rainforest/vines/8983/index.html
www.angelfire.com/journal2/chinfahshin/
introduction/intro.html
Chin Fah Shin

Learn About Butterflies – The Complete Guide
to the World of Butterflies and Moths
www.learnaboutbutterflies.com
Adrian Hoskins

SamuiButterflies
www.samuibutterflies.com
Leslie Day

Useful Contacts

Malaysian Nature Society
JKR 641, Jalan Kelantan
Bukit Persekutuan
50480 Kuala Lumpur
Malaysia
www.mns.org.my

Nature Society (Singapore)
510 Geylang Road
#02-05, The Sunflower
Singapore 389466
www.nss.org.sg

Penang Butterfly Farm
Jalan Teluk Bahang
11050 Penang
Pulau Pinang
Malaysia
www.butterfly-insect.com

Siam Society
131 Asoke Montri Road
(Sukhumvit 21)
Bangkok 10110
Thailand
www.siam-society.org

Siam Insect-Zoo and Museum
23/4 Mae Rim-Samoeng Rd
Mou 1, Tambon Mae Raem
Amphor Mae Rim
Chiang Mai 50180
Thailand
www.malaeng.com/blog/index.php

Photo Credits for Main Descriptions

Photos are denoted by a page number followed by t (top), b (bottom), u (upper), w (lower), m (middle), l (left) or r (right).

Anthony Wong: 29mr, 29br, 33t, 37t, 38t, 49t, 55b, 70b, 111tl, 114tl, 116tr, 130ml. **Antonio Giudici** (www.antoniogiudici.com): 9, 12tl, 24b, 25t, 27t, 30t, 34t, 35t, 39tr, 50b, 56t, 57bl, 60b, 61t, 62b, 65b, 68t, 77tr, 77umr, 77l, 79b, 81b, 84t, 86t, 89bl, 97br, 98bl, 99t, 100b, 102bl, 112t, 119bl, 121t, 124t, 126br, 129t, 137t, 142b, 144t, 146t, 150t, 150b, 154tl, 157m, 162b. **Benedict Tay:** 40b, 53t, 53b, 131b, 160b. **Benjamin Yam:** 28t, 44b, 54b, 66t, 83bl, 87tl, 98t, 110b, 115b, 116tl, 118b, 124m, 124b, 138t. **Bobby Mun:** 142t. **Chng Chuen Kiong:** 59t, 73b, 76bl, 84b, 96b, 115tr, 136br, 148t, 157t. **Dave Sargeant:** 42t, 42umr, 88b, 91b, 92b, 104bl, 137b. **Ellen Tan:** 13, 70t, 78b, 85bl, 95bl, 102tr, 102br, 103br, 106t, 121br, 147b, 149t. **Federick Ho:** 61b, 64br, 89tl, 99b, 115tl, 123ml, 140t, 140b, 143b, 158t, 158ml, 160t, 162m. **Games Punjapa Phetsri:** 72b, 101bl, 121bl, 125t. **Gan Cheong Weei:** 23b, 41t, 48b, 66b, 74b, 101br. **Glenn Q. Bagnas:** 100t, 101m, 108bl, 147m. **Goh Lai Chong:** 14t, 14b, 29t, 29ml, 30b, 39bl, 42wm, 42b, 43t, 50t, 57br, 63b, 67t, 72t, 75b, 76br, 78t, 80tr, 82br, 89tr, 93wm, 95t, 95br, 102tl, 103bl, 104t, 105t, 116rm, 117tr, 117tl, 118t, 120t, 122b, 127mr, 129br, 131tr, 133m, 133b, 134b, 135t, 136t, 138b, 139b, 149b, 151tl, 151mr, 151bl, 153tl, 153t. **Guek Hock Ping:** 107tl. **Hiroyuki Hashimoto:** 114bl. **Horace Tan:** 79tl, 88tr, 122tr. **James Chia:** 22b, 28bl, 40m, 81t, 87tr, 91t, 113b, 114tr, 130bl, 131tl. **John Moore:** 59m, 67b, 104br, 129bl, 139t. **Jonathan Soong:** 52b, 54t, 85tl, 112b, 113t, 123tr, 130tl, 144b, 146b. **Khew Sin Khoon:** 11, 12br, 29bl, 32b, 35b, 36t, 43m, 43b, 44t, 46t, 49b, 51t, 51b, 55t, 62t, 65tr, 65tl, 71t, 77br, 80bl, 82t, 87bl, 90tr, 90bl, 92t, 96t, 101t, 105br, 107tr, 109b, 120umr, 126ml, 126bl, 145t, 147t, 155b, 157b, 161tr, 161mr, 161b. **Koh Cher Hern:** 56b, 90tl, 132bl, 133t, 152t, 162t. **Laurence G. Kirton:** 17, 23t, 48t, 48m, 109tr, 109tl, 137m. **Leslie Day:** 97bl, 114mr, 128tl, 128tr, 161tl. **Liew Nyok Lin:** 24t, 80tl, 97tr, 103mr, 103tl. **Loke Peng Fai:** 45b, 64t, 64bl, 69t, 93umr, 108tr, 125br. **Mark Wong:** 83t, 117b, 127tr, 127br. **Mohd Walad Jamaludin:** 158bl, 158br. **Nawang Bhutia:** 58b. **Nelson Ong:** 27m, 36b, 55tr, 60t, 63t, 71b, 75t, 93t, 93uml, 111tr, 111b, 128bl, 148b. **Oleg Kosterin:** 39tl, 42uml, 154tr. **Peter Eeles:** 85tr, 154b. **Richard Ong:** 53m, 80b, 88tl, 108bm, 119tl, 143t. **Simon Sng:** 74t, 90br, 156b. **Steeve Collard:** 31t. **Sum Chee Meng:** 119br. **Sunny Chir:** 16, 22t, 26b, 28br, 37b, 47b, 52t, 68b, 69b, 83br, 86m, 86b, 105bl, 106b, 107br, 108tl, 108br, 110t, 114br, 120br, 123bl, 123br, 126tl, 126mr, 127bl, 130tr, 132br, 134t, 136bl, 141t, 152b, 155t, 159b. **Tan Ben Jin:** 39br, 79tr, 89br, 120uml. **Tan Chung Pheng:** 98m. **Tarun Karmakar:** 31b. **Tea Yi Kai:** 12bl, 25b, 26t, 27b, 32t, 33b, 38b, 40t, 41b, 45t, 46b, 47t, 51m, 57t, 73t, 76tl, 76tr, 77wmr, 82bl, 85br, 87br, 93bl, 93br, 94t, 94b, 97tl, 98br, 107bl, 116bl, 116br, 119tr, 122tl, 123mr, 125bl, 128br, 132tl, 132tr, 135b, 145b, 149m, 151tr, 153b, 156t, 159t, 159m, 160m. **Terry Ong:** 120wm. **Thawat Tanhai:** 12tr, 34b, 58t, 59b, 141b.

172